Archaeology and Bioarchaeology of Anatomical Dissection at a Nineteenth-Century Army Hospital in San Francisco

Bioarchaeological Interpretations of the Human Past: Local, Regional, and Global Perspectives

Archaeology and Bioarchaeology of Anatomical Dissection at a Nineteenth-Century Army Hospital in San Francisco

EDITED BY

P. Willey, Peter Gavette, Eric J. Bartelink, and Colleen F. Milligan

Foreword by Kenneth C. Nystrom

Series Foreword by Clark Spencer Larsen

University of Florida Press

Gainesville

University of Florida Press, 2023
Published in the United States of America

28 27 26 25 24 23 6 5 4 3 2 1

Library of Congress Cataloging-in-Publication Data
Names: Willey, Patrick S., editor. | Gavette, Peter, editor. | Bartelink, Eric J., editor. | Milligan, Colleen F., editor.
Title: Archaeology and bioarchaeology of anatomical dissection at a nineteenth-century army hospital in San Francisco / edited by P. Willey, Peter Gavette, Eric J. Bartelink, and Colleen F. Milligan.
Other titles: Bioarchaeological interpretations of the human past.
Description: 1. | Gainesville : University of Florida Press, 2023. | Series: Bioarchaeological interpretations of the human past: local, regional, and global perspectives | Includes bibliographical references and index.
Identifiers: LCCN 2022044602 (print) | LCCN 2022044603 (ebook) | ISBN 9781683402664 (hardback) | ISBN 9781683403487 (pdf)
Subjects: LCSH: Human dissection—California—San Francisco—History—19th century. | Hospitals—California—San Francisco—History—19th century. | Soldiers—Services for—California—San Francisco—19th century. | San Francisco (Calif.)—History—19th century—Hospitals. | BISAC: SOCIAL SCIENCE / Archaeology | MEDICAL / History
Classification: LCC QM33.4 .A73 2023 (print) | LCC QM33.4 (ebook) | DDC 611.009794/61—dc23/eng/20221007
LC record available at https://lccn.loc.gov/2022044602
LC ebook record available at https://lccn.loc.gov/2022044603

University of Florida Press
2046 NE Waldo Road
Suite 2100
Gainesville, FL 32609
http://upress.ufl.edu

UF PRESS

UNIVERSITY
OF FLORIDA

To those people whose bodies were used for medical training to advance medical practice and whose names have been lost to history

Contents

Figures

Maps

Tables

Appendixes

Supplemental appendixes are hosted by the University of Florida Digital Collections. All are freely accessible here: https://ufdc.ufl.edu/title-sets/AA00090501.

Appendix A. Supplemental Figures

Appendix B. Supplemental Tables

Foreword

The manner in which biological anthropology approaches the analysis of anatomized skeletal remains—remains that exhibit evidence of medical intervention such as autopsies, dissections, and amputations—has changed significantly.

Over the last decade, there has been a proliferation of reports and publications discussing skeletal remains recovered from historic contexts that exhibit evidence of medical intervention, with over a dozen journal articles and book chapters as well as two full length edited volumes. Examples have been reported on and discussed from Great Britain, Poland, France, Western Europe, Canada, and the US. This research can generally be placed into two, often overlapping, approaches; viewing the remains through the lens of medical history or social bioarchaeology.

Many of the earliest publications on anatomized remains were principally descriptive (e.g., Angel et al. 1987; Molleson and Cox 1993) or focused more exclusively on how this type of evidence informs medical history (Henderson et al. 1996; Hillson et al. 1998; Mann et al. 1991; Owsley 1991, 1995; Valentin 1995; Waldron and Rogers 1988; Wesolowsky 1991). At this point in time, most of the examples that had been recovered were isolated, with only one or two individuals within a larger skeletal collection exhibiting such evidence. Therefore, the description and discussion of such evidence often only occupied a small portion of the report. For instance, Angel et al. (1987) provide only a brief note about a craniotomy observed in the skeletal collection recovered from the First African Baptist Church in Philadelphia. Such remains were rarities and while providing a unique glimpse into the development of medical training, authors were largely not considering the remains in their broader sociopolitical context.

During the 1990s, research was published on two skeletal collections in which evidence of anatomization was widespread. The first collection was recovered during excavation of a cemetery associated with the Charity

Hospital in New Orleans. In his analysis and discussion, Doug Owsley (1995) focused on the historical context of medical training in the US and the role that Charity Hospital played. In his conclusions, however, Owsley touched on the potential social significance of the disproportionate number of African Americans remains that exhibited postmortem cutmarks.

The other large collection of anatomized remains analyzed during this time was discovered in the basement of the Medical College of Georgia. The research program that developed around the remains explicitly adopted a biocultural approach. This resulting publication was also notable because it represents the first book length treatment of anatomized remains—Robert Blakely and Judith Harrington's edited volume *Bones in the Basement: Postmortem Racism in Nineteenth-Century Medical Training.* In the volume, authors covered every subject, from the discussion of material culture to biological analyses (e.g., cutmark analysis), as well as the sociopolitical context of race and medicine in the nineteenth century.

I am making particular note of these features because in many ways this volume on the Point San Jose site is reflective of the one by Blakely and Harrington. Like its predecessor, it is an exhaustive analysis, utilizing a broad methodological tool kit to tease apart what this collection of bones can tell us. The volume covers the contextualized analysis of faunal bones, material culture, demographics, health and stress, mobility, and diet.

Contexts in which anatomized remains are found are deeply powerful. They often represent clandestine or illicit behavior and offer a stark picture on the treatment and disposal of human remains in the past. Their significance is not limited to telling us about past behavior, however, as their existence offers a chance to engage in a critical self-reflection.

More attention is being paid to the nature of the large skeletal collections in the United States—the George S. Huntington Anatomical Collection, the Robert J. Terry Anatomical Collection, the Hamann–Todd Osteological Collection, and the W. Montague Cobb Human Skeletal Collection. Akin to the research on anatomized remains recovered from archaeological contexts that have employed an interpretive framework grounded in the concept of structural violence, researchers have contextualized these documented collections from the same critical perspective—that the anatomization of black, brown, and poor bodies represents a form of institutionalized inequality (de la Cova 2019; Lans 2018; Muller et al. 2017; Watkins 2018).

There are still deeper connections. Even as we are reconstructing the behavior of nineteenth-century medical professionals and how they benefited from the exploitation of colored bodies, we have to acknowledge that we too are benefiting from the study of the same bones, often employing osteological methods derived from research with the anatomized remains of the very same marginalized groups. Here, Rachel Watkins draws a direct link between these bodies and inequities in the epistemological structure of the discipline itself wherein "there is a longer history of a critical mass of people of color being involved in bioarchaeological research as subjects (in the form of the skeletal data studied) than as knowledge producers" (Watkins 2020:20).

Rather than simply rarities that may sharpen our perception of the past, these remains speak to the treatment of bodies in the present—the bodies we study and the bodies of our colleagues. As these types of contexts and remains will no doubt continue to be uncovered, analyzed, interpreted, and published, it is imperative to think deeply not only about what they can tell us regarding the past, but also how they inform on our own positionality and the structure and future of the discipline.

Kenneth C. Nystrom

References Cited

de la Cova, C.

2019 Marginalized Bodies and the Construction of the Robert J. Terry Anatomical Skeletal Collection: A Promised Land Lost. In *Bioarchaeology of Marginalized People,* edited by Madeline Mant and Alyson Holland, pp. 133–155. Elsevier, Amsterdam.

Lans, A.

2018 "Whatever Was Once Associated with him, Continues to Bear his Stamp": Articulating and Dissecting George S. Huntington and His Anatomical Collection. In *Bioarchaeological Analyses and Bodies: New Ways of Knowing Anatomical and Skeletal Collections,* edited by P. K. Stone, pp. 11–26. Springer, New York.

Muller, J., K. Pearlstein, and C. de la Cova

2017 Dissection and Documented Skeletal Collections: Embodiments of Legalized Inequality. In *The Bioarchaeology of Dissection and Autopsy in the United States,* edited by K. C. Nystrom, pp. 185–201. Springer, New York.

Watkins, R.

2018 Anatomical Collections as the Anthropological Other: Some Considerations. In *Bioarchaeological Analyses and Bodies: New Ways of Knowing Anatomical and Skeletal Collections,* edited by P. K. Stone, pp. 27–47. Springer, New York.

Watkins, R. J.

2020 An Alter(ed) Native Perspective on Historical Bioarchaeology. *Historical Archaeology* 54(1):17–33.

Series Foreword

Human remains are derived from a remarkable range of archaeological contexts. The greater volume of human remains derived from these settings comes from cemeteries representing deceased members of specific communities. There are, however, many exceptions to these kinds of discoveries and follow-up investigations. Historical medical facilities are among the most extraordinary bioarchaeological settings. In this regard, this volume presents a novel context, namely a series of commingled remains representing individuals associated with a nineteenth-century hospital at Point San Jose, California. The archaeological recovery process revealed the presence of largely disarticulated, commingled skeletal remains. Based on the survey of skeletal and dental elements present, the series represents at least 25 individuals derived from anatomical dissections undertaken from an associated hospital. The bioarchaeological research team headed by P. Willey provides details on the series and addresses key questions relating to purpose, context, and outcome. The analysis of the remains is put in the larger historical context of late-nineteenth century medical practice involving dissection. The deposit of remains in this setting is from a time when emerging medical schools were keenly interested in having access to bodies of deceased individuals for teaching anatomy and preparation of future physicians. Importantly, the bioarchaeological study of the remains provides a window onto practices not just geared at dissection, but rather, training of physicians at a time when there were so few of them in the frontier West, including California.

The recovery and study of the Point San Jose remains provided the chance to develop a larger narrative of the beginning of medical teaching and practice in a frontier context. In particular, the discussion of this specific context provides a rich picture of taphonomy and the associated modifications in the skeletal remains by environmental, animal, and human sources. Importantly, the book reveals considerable details on medical practice in the second half of the nineteenth century, including manner of treatment as viewed from cutmarks and other evidence of dismemberment, preferential identification, and selection of specific skeletal elements prior to disposal. For example, femoral and cranial elements are less represented than other skeletal elements. These findings suggest that the medical practitioners heading the dissections had selected femoral and cranial elements saved for curation and later study.

Among the most compelling discussions in the volume is the demographic record and dietary inferences giving insights into the lives of the persons represented by the Point San Jose skeletal remains. Not surprisingly, the community and individuals represented in the Point San Jose setting focused their diets on local resources, largely based on farming and marine resources available in the San Francisco Bay Area. In this regard, the analysis revealed both for this setting and the surrounding population in general a homogeneous dietary regime. Whatever these foods may have been, the remarkably poor oral health as represented by elevated dental caries prevalence supports the notion that plant carbohydrates provided an important part of daily diet.

Building on a series of questions guiding their research, the contributors to the book give fundamental insights into the lives and health of the individuals represented in this setting. As so carefully pointed out, the skeletal series studied by P. Willey and collaborators is not necessarily represented by the larger society living in the region in the nineteenth century. However, the study of the series of skeletons provides important insights into the health and lifeways. Importantly, the analysis gives perspective on medical teaching practices in a remarkable setting of settlement of nineteenth century California.

Clark Spencer Larsen
Series Editor

Preface and Acknowledgments

On the afternoon of Monday, October 25, 2010, archaeologist Tim Spillane made a startling discovery outside a historical building at Fort Mason, San Francisco, which was in the process of being renovated. The building is part of Golden Gate National Recreation Area (GGNRA), a unit of the National Park Service (NPS). While monitoring the excavation and remediation of lead-contaminated soils around the perimeter of the building, a cast iron plate was discovered and lifted, revealing human bones, faunal remains, and other artifacts. Spillane, following NPS protocol, contacted authorities, and the San Francisco Office of the Chief Medical Examiner dispatched Chuck Cecil, a forensic anthropologist, to the site. Cecil concluded that this discovery was not a crime scene and fell outside his purview, so the medical examiner relinquished jurisdiction to GGNRA personnel.

Leo Barker, the GGNRA park archaeologist who had contracted Spillane to monitor the remediation work, assembled a crew to excavate the rest of the bones. The crew consisted of three archaeologists (Spillane, himself, and another GGNRA archeologist, Peter Gavette). The following morning, Spillane described the previous day's events, then the crew commenced the excavation. NPS Park Ranger Skot Jonz joined the archaeologists, as well as NPS cultural resources employees Abbie-Sue Fisher and Paul Scolari, and NPS Realty Specialist Catherine Robles. Because some of the long bones recovered the previous day showed saw marks, and the fact that the adjacent building was once the military post hospital, they initially believed the bones represented amputations. More long bones were recovered early that second day of excavations, as well as smaller bones, including a sacrum. That bone proved that this pit did not contain medical waste resulting from the amputation of limbs alone. That week, 18 additional people assisted the archeologists. They included Wesley Barker, Wes's friend, Lulu Chye, Bill Clevenger, Kate Clevenger, Abby Sue Fisher, Stephen Haller, Bob Holloway, Scott Jonz, Justine Lai, Catherine Robles, Paul Scolari, Seren Sehota,

Aurora Tucker, as well as another unidentified woman. Following the fieldwork, Angela Locke Barton and Bill Clevenger, a physician, began an initial assessment of the human remains. In 2015 the Golden Gate National Parks Conservancy (GGNPC) contracted Locke Barton to catalogue and analyze the artifacts.

California State University, Chico (Chico State) became interested in the Point San Jose human bones when Thomas Lowry, then living in Woodbridge, Virginia, mailed P. Willey a newspaper clipping. According to the article, bones, buttons, and bottles had been discovered "within view of the Golden Gate Bridge" (Fagan 2010). Interest piqued, Willey contacted Jonz, who had been mentioned in the newspaper article. In March 2011, Eric Bartelink, Colleen Milligan, and Willey from Chico State scheduled a visit to the GGNRA's archaeology laboratory near Rodeo Beach, Fort Cronkhite, to inspect the bones that Barker curated there. As follow-up, they recommended that Barker pursue having the human bones analyzed and provided an outline of what those analyses should include.

In August 2015, a year before his anticipated retirement, Barker contacted Bartelink, Milligan, and Willey at Chico State to begin the osteological portion of the project. Perceptively Barker established two phases to the osteological work. The first one involved basic processing and descriptions, followed by a second one, which delved into more depth and pursued more specialized studies.

While Chico State and GGNRA personnel juggled the logistical arrangements for the human bones, personnel at both institutions processed the contract. For both stages of the project, Julie Jessen at Chico State's Research and Sponsored Programs (RESP) focused the process, and helped develop a budget and scope of work. On the NPS's part, financial support came from a regional level cultural resource fund, which was channeled through the GGNPC. On the behalf of GGNPC, Garrett Lee, Projects Finance Manager, provided information and logistical know-how. As the proposal and budget matured, RESP's John Miner assumed our project and finally at RESP, day-to-day project oversight fell to Tara Becraft.

On April 13, 2016, Barker and Peter Gavette, GGNRA archaeologists, transferred the Point San Jose bones to Chico. Starting a few months later, Chico State graduate student Maria Cox ordered project equipment, completed hiring paperwork, arranged new-hire orientations, and organized the summer's laboratory work. Student workers consisted of nearly all of the then-active graduate students enrolled in the physical anthropology subdiscipline at Chico State. In addition to Cox, they included (in alphabetical order) Matt Bond, Kristen Broehl, Jessica Curry, Martha Diaz, Sarah Hall, Kelsie Mae Hart, Heather MacInnes, and Valerie Sgheiza. In addition, the Human Identification Laboratory Supervisor Alex Perrone oversaw portions of the work.

To them fell the tedious labor of identifying, matching, cataloging, reassembling, and photographing the Point San Jose materials. They inventoried and labeled the Point San Jose human elements as well as measured the bones, estimated age, sex, and ancestry. The three principal investigators checked identifications and assessments. Once those preliminary tasks were completed, pilot osteological analyses were conducted and draft sections written.

On August 15, 2016, the GGNRA archaeologists and Chico State osteologists met at Fort Cronkhite for a Point San Jose seminar. There, participants presented archaeological, historical, and osteological findings. Barker retired at the end of that month and Gavette succeeded him. A draft first-phase osteological report was submitted to GGNRA authorities. Following review and recommendations, the draft was corrected and a report accepted by GGNRA authorities in September 2016 (Willey et al. 2016).

With conclusion of the preliminary descriptive work, funding for continuation of the project was negotiated. While Chico State and GGNRA personnel discussed topics requiring additional research, the same people at GGNPC and RESP who managed the first contract negotiated additional contractual obligations. A continuation contract was signed in January 2017 to extend the Point San Jose osteological work through August 2018.

The second osteological phase emphasized seven topics, refining earlier assessments and extending efforts to additional specialized topics. Advanced osteological topics involved the human elements and number of individuals represented, paleodemographic comparisons and reassessments, refined paleopathological assessments and expanded radiography, and detailed study of taphonomic alterations including X-ray fluorescence (XRF) of selected stains. Additional topics initiated in this second stage involved assessment of archaeological context of materials in the pit, identification of faunal elements and their modifications, and isotopic assessment to reconstruct diet and estimate region of origin.

Many of the same Chico State students who participated in the preliminary osteological analyses continued their work in the second stage, with a few newcomers. In addition to the people listed above, Sam Mijal joined the work. Cox again oversaw day-to-day administrative logistics until she left for further graduate study, then Mallory Peters took over those on-location responsibilities.

Chico State graduate students Kasey Cole and Hart initiated an assessment of faunal elements, identifying the species represented and bone alterations as well as interpreting them. Cox and Sgheiza continued exploring human element representation and number of individuals represented. As part of this topic, Bartelink and coworkers attempted to re-associate commingled portions of individuals using XRF. Sgheiza and Willey re-evaluated the paleodemographic profile. Using stable isotopes, Bartelink and Hall estimated the remains' regions of origin. Another isotopic study by Hall, Bartelink, and Julia Prince-Buitenhuys reconstructed diet. In paleopathology, Milligan, Broehl, Hart, Karin Wells, Curry, Peters, Vanessa Reeves, and Bartelink expanded and elaborated on studies of linear enamel hypoplasia and degenerative joint disease, and provided more detailed assessment of ribs and vertebrae. Bartelink, Curry, and Peters documented saw and incision cut marks, emphasizing cut mark distribution by element and direction, as well as attempting to identify chemical preservatives and staining. The second phase report was completed in late July 2018 (Willey et al. 2018).

A culminating event occurred, which was devoted to the Point San Jose discovery, analyses and interpretations. In April 2018, Chico State and GGNRA personnel as well as some other specialists participated in a symposium at the Society for American Archaeology's annual meeting in Washington, DC. In addition to that symposium, supplementary papers and posters have been presented at other meetings, including the Society for California Archaeology and the Paleopathological Association. Three theses based in part on the Point San Jose human skeletal remains have been completed (Broehl 2018; Hart 2019; MacInnes 2017).

In addition to the earlier efforts listed above, other participants have joined us to produce this book. They are presented in approximate chronological order. Kristen E. Pearlstein, Collections Manager, and Brian Spatola, Anatomical Collections Curator, Anatomical Division, National Museum of Health and Medicine, have followed this project since its inception. Although they were unable to participate in the culminating symposium, they have provided a chapter on the post–Civil War surgeons, the Army Medical Museum, and military medicine for this book.

When additional Point San Jose faunal materials were discovered in 2019, Chico State graduate students Cecily Merwin and Alina Tichinin rose to the challenge, analyzed the additional remains, and collaborated with the zooarchaeologists who conducted the initial assessment.

Chico State alumna Lisa Bright came to the project late. In the brief time available, she provided a historical overview of Bay Area history as well as key developments in the military history of San Francisco.

When in May 2019 publication outlets were considered, we sought advice from Debra L. Martin, University of Nevada, Las Vegas. She generously weighed our options and recommended the University of Florida Press (UFP). Following her suggestion, we forwarded a draft proposal to Clark Larsen, Ohio State University, editor of UFP's bioarchaeology series. Three anonymous peer "readers" reviewed a draft of this work and made valuable suggestions. As the manuscript neared publication, Marthe Walters and Nick Walther provided editorial direction.

Throughout this project, librarians and staff members have aided our research and queries. In particular, Flora Nunn, Interlibrary Loan Agent at Chico State's Meriam Library and staff members at the University of Minnesota's Owen H. Wangensteen Historical Library of Biology and Medicine come to mind. David Philhour and Lee Brisbane-Ryan opened Chico State's Behavioral and Social Sciences' Statistical Laboratory doors, normally locked during the 2016 summer. Michael Magliari, a Chico State historian, provided bibliographical sources. Jeanette R. Methvin-Terry, Chico State Anthropology Department Coordinator, oversaw the daily requests at the departmental level. To all these people and many others, we extend our thanks and gratitude. We hope this listing shows how monumental a research undertaking as this is and how many individuals it requires to guide it to fruition.

References Cited

Broehl, Kristen

2018 Osteometric Sorting of Metacarpals and Metatarsals in Commingled Human Skeletal Series, Master's thesis, Anthropology Department, California University, Chico.

Fagan, Kevin

2010 Fort Mason Yields Historical Treasure. *San Francisco Chronicle* 3 November:C1, C8. San Francisco, California.

Hart, Kelsie M.

2019 Identifying Autopsy and Dissection on Human Skeletal Remains: A Case Study from Point San Jose (Fort Mason), Golden Gate National Recreation Area, Master's thesis, Anthropology Department, California University, Chico.

MacInnes, Heather

2017 Specimens of Morbid Anatomy: A Study of Context and Sorting Methods of Commingled Human Remains from the Point San Jose Ossuary, Master's thesis, Anthropology Department, California University, Chico.

Willey, P., Colleen Milligan, Eric Bartelink, Maria Cox, Kristen Broehl, Jessica Curry, and Valerie Sgheiza

2016 *Descriptive Analysis of Human Skeletal Material from Point San Jose (Fort Mason), Golden Gate National Recreation Area.* Submitted to Golden Gate National Recreation Area Conservancy, Proposal 16–131.

Willey, P., Eric Bartelink, Colleen Milligan, Maria Cox, Mallory Peters, Kasey Cole, Jessica Curry, Sarah Hall, Kelsie Hart, Sam Mijal, Julia Prince-Buitenhuys, and Valerie Sgheiza

2018 *Osteological Analyses of a Medical Waste Pit from Point San Jose (Fort Mason), Golden Gate National Recreation Area.* Submitted to Golden Gate National Recreation Area Conservancy, Proposal 16–131.

Introduction, Comparable Skeletal Series, and Organization of Volume

P. WILLEY, ERIC J. BARTELINK, PETER GAVETTE, AND COLLEEN F. MILLIGAN

No visit to San Francisco is complete without visiting the Golden Gate Bridge and seeing Alcatraz Island (Map 0.1). Near the northern end of the city, the Golden Gate Bridge, which opened to traffic in 1937, commands an iconic, looming presence. About three miles east of Golden Gate, Alcatraz Island provides a diametrical counterpoint. It is dreary, dilapidated, dank, and eroded.

After its service as a military post, Alcatraz served as a military prison (1861–1933) and later as a federal penitentiary (1933–1963) for the most notorious and intractable prisoners. Infamous convicts who did time there included Robert "Birdman of Alcatraz" Stroud, Al Capone, and George "Machine Gun" Kelly.

Within sight of both the Golden Gate Bridge and Alcatraz is Point San Jose. On firmer, though still earthquake-prone ground, Point San Jose is now part of Fort Mason National Historic District and more broadly the Golden Gate National Recreation Area. Remarkable events had happened at Point San Jose. But unlike those two iconic landmarks just mentioned, Point San Jose's secrets had been buried soon after Alcatraz received its first military prisoner and long before the first car crossed Golden Gate Bridge. The secrets of Point San Jose went unknown for 150 years.

That secret surfaced on October 25, 2010. While monitoring lead remediation efforts in the soils around the former Point San Jose hospital (1864–1903), archaeologists discovered a pit yielding artifacts and commingled, mostly disarticulated human remains. Excavation of the feature revealed a plethora of waste material presumed to be from the hospital in addition to

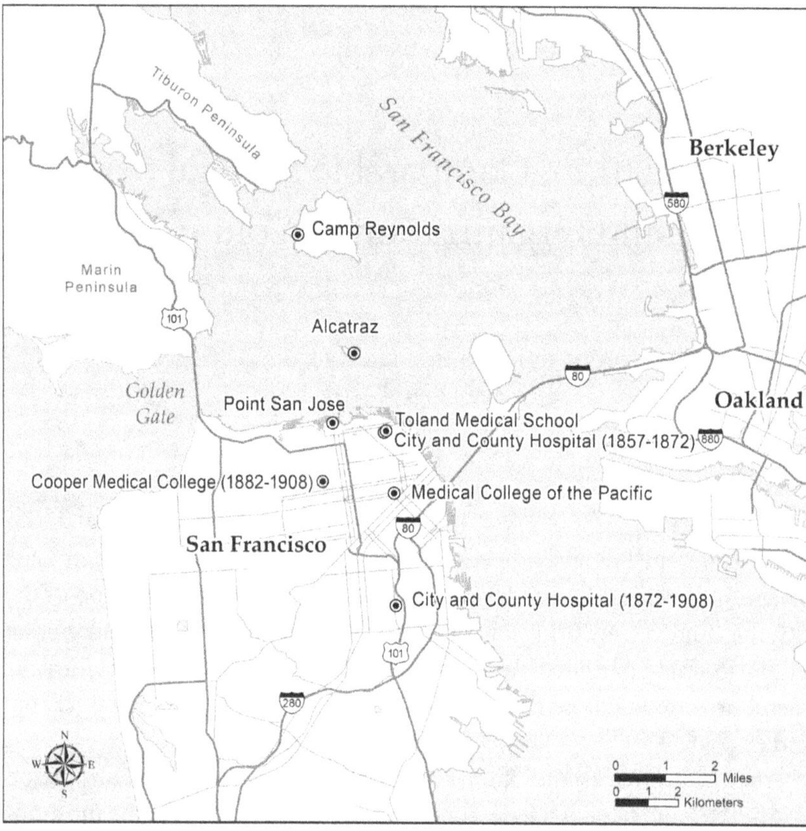

Map 0.1. San Francisco Bay Area showing Point San Jose and other locations mentioned in the text. Created by Judy Stolen.

the skeletal elements. Dating ca. 1860–1900, the human remains consisted of 4,450 commingled elements, representing approximately 25 individuals. Many of the bones showed the results of anatomical dissection. Following common practice in nineteenth-century cadaver acquisition, they likely represented the remains of people representing poor and marginalized groups in the Bay Area.

The pit and its contents have been the focus of an extensive research project since their 2010 discovery. On its discovery, initial questions arose and additional ones have emerged during the last decade. These general questions have guided our investigations.

> What purpose had the discarded remains served? Did they represent a pathological collection, cadavers, autopsied corpses, or members of the military or Native American groups?

When were the bones and artifacts buried in the Point San Jose pit and what did their disposal indicate?

Did the Point San Jose human bones include all elements of all skeletons? Or did they represent selected skeletal elements and chosen anatomical segments?

What is the demographic composition of the human remains? What sex, age, and ancestry groups do the Point San Jose skeletons represent? Were these individuals local to the San Francisco Bay Area or nonlocal?

What do the human remains indicate about their health and stress during life?

On a broader level, what role did the Point San Jose deposit represent in post–Civil War medicine and military medicine? And more specifically, who commanded gathering, use and disposal of the human remains?

Here, we embrace a holistic perspective on the discovery and contents of the pit to answer those questions. The deposit offers extraordinary perspectives on late nineteenth-century medical understanding, knowledge, practices, and education as well as the cultural context in which bodies were acquired for dissection. We use historical, archaeological, and osteological approaches to describe, document, and interpret the Point San Jose pit.

Similar North American Historical Osteological Series

Although extraordinary, the Point San Jose discovery and its materials are not unique. Similar finds have been made elsewhere and many of those unearthings have been described in print. In fact over the last decade, several syntheses have been published that present and summarize the importance and interpretations of such skeletal series (Nystrom 2014, 2017; Stone 2018). Only within that context of previously published, comparable sites can the Point San Jose discovery be understood.

Those comparable studies describe bones with indications of medical interventions (amputation, etc.), dissection, and autopsy. Before presenting the similar series, we need to consider those three processes and their alterations as well as the intentions behind them.

Medical interventions involve living people. The goals of medical interventions include restoring health, relieving suffering, and preserving life. As an example of medical intervention, amputation consists of the surgical removal of a portion of a damaged anatomical structure. Without amputation the impaired body portion may jeopardize the person's life.

Dissections involve cadavers. Dissection procedures include slicing away tissues to reveal underlying, internal structures, and anatomical organization. The goals of dissections are to study anatomical structures, systems, and organs. The work in an anatomical laboratory provides an example of dissection.

Autopsy also involves a corpse. Similar to dissection, it exposes internal structures for study; however, the goals of these two procedures differ. An autopsy's purpose is to find, identify, and to interpret the cause and manner of a person's death. Cause of death comprises the underlying cause of a person's demise. For example, cardiovascular disease (e.g., heart attack, stroke, heart failure) is the most frequent cause of death in the modern US (Murphy et al., 2018). Manner of death, on the other hand, indicates the broad categories of death. Generally, manner of deaths consist of natural, accidental, suicide, homicide, and undetermined categories.

The three processes leave similar osteological marks, so all three are considered here when summarizing similar North American discoveries. Before tackling that challenge, the similar processes and sites that are not reviewed here should be noted.

Reports of many similar modified skeletons come from western Europe (e.g., Anderson 2002; Boston and Webb 2012; Bugaj, et al. 2013; Chamberlain 2012; Charlier, et al. 2013; Fowler and Powers 2012; Henderson et al. 1996; Hillson et al. 1998; Hull et al. 2003; Valentin and d'Errico 1995; Western 2012; Western and Bekvalac 2015). We omit those studies from this consideration because medical education and practice there differed from North America. This discussion also excludes North American prehistoric archaeological materials (see, e.g., Larsen 2015:115–177; Cook and Powell 2006:314–315). It is difficult to determine when in the perimortem interval those prehistoric alterations occurred, and their often ambiguous archaeological contexts make the underlying purposes uncertain. We also exclude more recent forensic contexts, such as criminal dismemberment and mutilation (e.g., Black et al. 2017; Ross and Cunha 2019).

Unfortunately, many reports of those modified human remains in North America are available only as personal communications, "gray" literature, abstracts from proceedings, personal communications, theses, dissertations, or in press. Most of those more obscure reports are omitted from this summary. We also omit curated anatomical collections from consideration. The two best-known anatomical series in the US are the Terry Anatomical

Collection and Hamann-Todd Osteological Collection. There is little mystery to the origin and function of such anatomical collections.

Here, we concentrate on modified human skeletal remains found in historical North American archaeological contexts. This review is mostly restricted to widely available, professionally published skeletal series. For additional comparable skeletal series, see Nystrom (2014, 2017b:5, table 1.1). In addition to those previously summarized studies, several recent discoveries have been published and are summarized here.

Among the better documented and published North American archaeological-historical skeletal series with modifications, two dozen studies shine. They are presented in approximate chronological sequence (Table 0.1). Rather than on a site-by-site basis, we summarize them here by time, location, burial context, demography, taphonomy, and purpose.

The modified skeletal series span the time spectrum for historical North America. They date from earliest colonial times (e.g., Champlain's Cemetery) to the mid-twentieth century (e.g., possibly the skeleton at University of North Dakota). Many discoveries come from the nineteenth century, with most of those being in the mid- or late nineteenth century. That date range includes the materials at Point San Jose. This predominance of nineteenth-century discoveries probably results from forgotten or overlooked properties that have become valuable recently for new developments.

Similar to the temporal span, the location of the discoveries also shows a wide but not random geographical distribution (Map 0.2). They spread from the eastern US (e.g., Spring Street Presbyterian Church, Manhattan, New York) to the western US (Legion of Honor, San Francisco), and north (Champlain's Cemetery, St. Croix Island, Maine) to south (Charity Hospital Cemeteries, New Orleans). Most of the altered skeletons, however, have been discovered in the eastern and southern US, and in cities and urban areas. This geographic distribution may be attributed to locations with denser populations where property values have risen and construction is more concentrated. With the exception of the Legion of Honor materials, the Point San Jose materials are the most westward of the altered remains. Similar to many other discoveries, Point San Jose is located in a large metropolitan area and discovered during construction-related operations.

Modified skeletal remains come from a variety of archaeological contexts—cemeteries and clandestine being the two most common circumstances. Many comparably modified skeletal remains originate in formal, sanctified cemeteries. These traditional cemeteries may be intended for the dead of a community (e.g., Alameda-Stone Cemetery), a congregation (e.g.,

Table 0.1. Summary of North American skeletons showing evidence of medical intervention, autopsy or dissection, in approximate chronological order

Site	Location	Date	Skeletons				Taphonomic alterations	Interpretation	References
			Age	Sex	Ancestry	Number of Individuals			
Champlain's Cemetery	St. Croix Island, Maine	1604–1605	Young adult	Male	European	1	Craniotomy	Autopsy	Crist et al. 2004, Crist and Sorg 2017
Scattered	Fort James, Virginia	1611–1617	Adolescent and adult	Male	European	3	Trephinations and craniotomy	Medical/anatomical specimens	Bruwelheide et al. 2017
Midden, Charlton's Coffeehouse	Williamsburg, Virginia	1760s	Adolescent and adult	—	—	2	Incisions and hand phalanx	Dissection and articulated hand	Chapman and Kostro 2017
Well, Medical College	Richmond, Virginia	1848–1860	At least 9 subadults (< 14yrs) and 44 adults (>15 yrs)	Ca. 1/3 females, ca. 2/3 males	2 European, 18 African heritage	>50	Sectioning, trephining, sawing, incising, and articulating or mounting	Dissection, and training in trepanations, amputations, and autopsies	Owsley et al. 2017
Snake Hill	Fort Erie, Ontario	War of 1812	Most young adults, >1 adolescent	Males	European descent	7	Sawed femora, humeri, radius and ulna	Amputations	Saunders 1991, Pfeiffer 1991, Owsley et al. 1991
Dry well, Holden Chapel	Harvard University, Cambridge, Massachusetts	1801–1850	1 fetus, 2 infants, and 1 child as well as 12 adults	Both sexes	—	>16	Sawed crania, ribs, sterna, femora, tibiae and fibulae. Vertebral sequences mounted	Anatomized specimens, surgical training	Hodge 2012, Hodge et al. 2017
Burial vaults, Spring St. Presbyterian Church	Manhattan, New York	1820–1850	infant, adolescent and middle-aged adult	Male?, male	Non-European	3	Sawed craniotomies	Dissection and autopsy	Novak and Willoughby 2010, Novak 2017
First African Baptist Church Cemetery	Philadelphia, Pennsylvania	1823–1841	Adult	Female	Black	1	Sawed cranial vault	Autopsy	Angel et al. 1987
Albany County Almshouse Cemetery	Albany, New York	1826–1926		15 females, 30 males		51	Craniotomies, small drill holes, trephination, transverse limb bone saw cuts, and cuts in some postcranial	Dissection	Lusignan 2004, Lowe 2017

Medical College	Augusta, Georgia	1830–early 1900s	43 >20 yrs, 5 < 20 yrs	>5 females, >17 males	>17 Blacks, >5 Whites		Severing or cut marks	Dissection, discarded cadavers	Blakely and Harrington 1997, Harrington and Blakely 1995a, 1995b, McFarlin and Wineski 1997
Charity Hospital Cemeteries	New Orleans, Louisiana	Most 1847 –1929	92% adults*	43% female, 55% male*	Two or three times as many Blacks as Whites*	271*	35–40% of crania sawed. Sawed mandibles	Amputation, surgery & autopsy possibility dissection	Owsley (1995)
Charity Hospital Cemeteries	New Orleans, Louisiana	1847–1929	—	—	—	—	Craniotomies, chest plate removal, incisions near muscles, sawed/sectioned mandible	Amputation/surgery, autopsy, dissection, experimentation,	Halling and Seidemann 2017
Newburgh Colored Burial Ground	New York, New York	Ca. 1830–1870	Young adult	Female	Black (based on context)		Craniotomy	Dissection or autopsy	Nystrom 2011
Blockley Almshouse Cemetery	Philadelphia, Pennsylvania	Ca. 1865–1895	Ca. ½ subadults, ½ adults	Ca. 1/3 females, 2/3 males	4% African, 1% Asian/Native American, 95% European	Ca. 248	Trephinations, craniotomies, sections, saw cuts, incisions, one healed amputation	Autopsies, dissections, instruction	Crist and Crist 2011, Crist et al. 2017
Fort Craig Post Cemetery	Rio Grande River, New Mexico	1854–1885	Adults	Males	Black and White	5	Craniotomies, laminectomies, other sawed bones	Autopsies, amputations, dissection	Goff 2009
Driveway construction	Sommerville [sic], Massachusetts	mid-1800s	Fetus, adolescent, young and middle-aged adults	>2 males	—	>6	Sawing and trephining	Dissection waste	Saul and Saul 1986a, 1986b, 1987
Poorhouse Cemetery	Erie County, New York	1850–1870	one child, others adult	>3 females/?, >9 males/?	—	20	Craniotomies, laminectomy, transection/sectioned long bones, and thoracotomy	Dissection and/or autopsy, amputation	Nystrom 2014, Nystrom et al. 2017c

(*continued*)

Table 0.1—*Continued*

Site	Location	Date	Skeletons				Taphonomic alterations	Interpretation	References
			Age	Sex	Ancestry	Number of Individuals			
Dunning Poorhouse Cemetery	Chicago, Illinois	Ca. 1851–1870	Adults	Both sexes	—	5	Transverse limb bone sections, craniotomy, drilled circular plugs	Dissection and medical practice	Grauer et al. 2017
Alameda-Stone Cemetery	Tucson, Arizona	ca. 1860–1881		>2 males	Euro-American, Hispanic	3	Removal of the top of skulls	Possible autopsies	Heilen et al. 2012
DeSaussure College	Columbia, South Carolina	1867–1876	Adults	>female, >male	Possible Asian-Native American	>3	Craniotomy, sawing, staining	Autopsy, dissection, demonstration/assemblage, pathology instruction	Stevens et al. 2018
Legion of Honor	San Francisco, California	1868–1906	Only adults selected for analysis	30% female 70% male	Not assessed	90 of 900 individuals analyzed	"amputated body parts, . . . preserved medical specimens, and . . . hospital waste."	Amputation, dissections?, and anatomical waste?	Buzon et al. 2005
Freedman's Cemetery	Dallas, Texas	1869–1907	Adults	Males	Black (based on context)	2	Sawed calvarium and femora	"Medical cadavers"	Davidson 2007
Physician's privy	Annapolis, Maryland	Late 1800s	Subadult (9–12 years), young or middle-aged adult	Female	—	>2	Sawed femora	Postmortem anatomical experimentation	Mann et al. 1991
Institution Grounds Cemetery #2	Milwaukee County, Wisconsin	1882–1925	Mostly middle-aged adults	Mostly males	—-	165	Craniotomies, trepanations, sectioned bones, incisions, saw cuts, and decapitation	Autopsies and dissections	Dougherty and Sullivan 2017, Richards et al. 2017
University of North Dakota campus	Grand Forks, North Dakota	1948	Middle-aged adult	Female	Black	1	Dismemberment, cuts on ribs and skull, adhering blue chalk	Dissected cadaver, anatomical waste	Stubblefield 2011

Source: Modified and elaborated from Nystrom 2017:5, table 1.1.

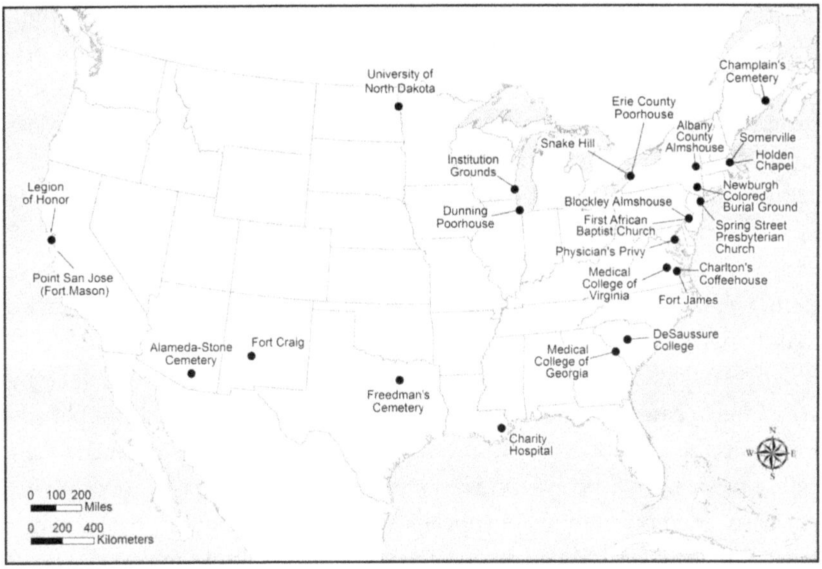

Map 0.2. Major sites in North America with skeletons showing similar modifications as those discovered at Point San Jose. Created by Judy Stolen.

Spring Street Presbyterian Church), racial or ethnic group (e.g., Newburgh Colored Burial Ground), military establishment (Fort Craig post cemetery), or some combination of those groups. Other cemeteries lie adjacent to institutions that served impoverished groups. For instance, they may be associated with almshouses (e.g., Blockley Almshouse Cemetery) or charity hospitals (e.g., Charity Hospital Cemetery).

In addition to remains buried in formal, set-aside cemeteries, modified skeletal elements also have been discovered in less formal locations. After serving their purposes, these remains appear to have been discarded with rubbish. Those discarded locations include middens (e.g., Fort James and Charlton's Coffeehouse), wells (e.g., Holden Chapel), basements (e.g., Medical College of Georgia), and specially prepared pits (De Saussure College, Somerville, and University of North Dakota). Point San Jose compares to the latter discoveries: a specially prepared pit in an obscure, unsanctified spot.

Moving beyond location and context, the paleodemographic makeup of altered skeletons shows both variety and selectivity. Paleodemographic variables include age, sex, ancestry, and numbers of individuals represented.

Ages of altered skeletons range from fetal, through childhood, adolescence, and into adulthood. Rather than a normal mortality distribution, most of the skeletons represent adults (cf. Medical College of Virginia). In contrast to the altered skeletons' mostly adult ages, more typical attritional mortality distributions show greater proportions of subadult deaths (Chamberlain 2006). Point San Jose has the spectrum of ages and a mortality curve that follows the trends at comparable sites, contrasting with normal mortality distributions.

Both sexes are represented in most of the other altered skeletal series, with males generally outnumbering females (e.g., Erie County Poorhouse Cemetery). Point San Jose follows this trend of greater proportions of males than females.

Ancestral affiliations of the altered skeletal series show diversity. Some are of European descent, others African, followed distantly in numbers by Hispanic and possibly Native American and Asian descent. In similar skeletal series when Europeans and Africans are recovered from the same site, Africans often outnumber Europeans by two or three times (e.g., Charity Hospital Cemetery). Similarly Point San Jose has multiple ancestral groups present, although in different proportions and heritages than previous osteological reports.

When investigating the number of altered individuals present, much variation occurs across the sites. Based on minimum number of individuals (MNI), some sites reveal only a single person (e.g., First African Baptist Church), while a few others soar into the hundreds (e.g., Blockley Almshouse Cemetery). Point San Jose's MNI of 25 probably falls near the modal number of altered individuals in each site.

Counting individuals and estimating their biological profiles often becomes complicated. Many of the altered skeletal series consist of single, relatively complete skeletons (e.g., Alameda-Stone Cemetery). Depending on skeletal completeness and preservation, osteological assessments may provide reliable counts and estimations of age, sex, and ancestry. But many of the skeletal series consist of commingled, partial, fragmentary, and incomplete remains (e.g., DeSaussure College). Point San Jose falls into that latter category of uncertainty.

The presence of postmortem skeletal modifications define these skeletons of interest and separate them from other series lacking such modifications. Modifications noted in previous publications include incisions, saw cuts, drilling, and mounting hardware. Those modifications have been interpreted as indications of trephination (e.g., Albany County Almshouse Cemetery), craniotomy (e.g., DeSaussure College), mounting for display (e.g., Medical College of Virginia), and dismemberment and decapitation (e.g., University of North Dakota campus).

Each of the three processes (amputations, dissections, and autopsies) has a distinctive, different purpose, but functions may overlap and skeletal alterations may merge. Differentiating dissection from autopsy has proved particularly challenging (Dittmar and Mitchell 2015). In addition to postmortem alterations, antemortem amputations and other surgical interventions may have occurred shortly before death and may appear skeletally consistent with postmortem alterations. Nystrom (2017b:4–8) notes the osteological challenges to diagnosing dissection, autopsy, and in some cases, these processes from amputation. Whether in the living, vital person or following death for some indefinite and variable period, fresh bone responds in similar fashion. Consequently, the modifier's intentions and plans can only be inferred, not known for certain. These intentions, while often vague and ambiguous, remain important considerations for understanding both medical and anatomical knowledge as well as the social context of the remains (Nystrom 2017b:8).

Although complex and confounding, assessing whether the Point San Jose bones represent amputations, dissections, and/or autopsies is a goal of the present work. Placing these alterations and the individuals in the context of anatomical/medical knowledge and social context are fundamental steps.

Although admittedly not unique, the deposit at Point San Jose remains extraordinary. It is a rare example of a western US autopsied or anatomized series. Only five other skeletal series hail from west of the Mississippi River (Fort Craig, New Mexico; Alameda-Stone Cemetery, Tucson; Freedman's Cemetery, Dallas; University of North Dakota campus, Grand Forks; Legion of Honor, San Francisco). Point San Jose is one of only two discoveries from the West Coast.

Point San Jose is extraordinary in another aspect. Almost all of the other series, both in the western US and elsewhere, have been associated with civilian activities, not military dealings. The other two military exceptions consist of Fort Craig and Snake Hill. Similar to them, the Point San Jose materials were found on what had been a military establishment and near the post hospital. This location combined with the deposit's date indicate a military association.

Organization of Volume

This present work follows a standard organizational format, divided into several broad sections. Here are section-by-section summaries of those chapters in the sequence that they appear in this work. For specific information on each chapter, see the introduction preceding each section.

History and Background

This section provides a historical context for the Point San Jose discovery. It includes a general history of the San Francisco Bay Area, background of nineteenth-century cadaver acquisition, roles of post–Civil War Army surgeons, and a biography of Edwin Bentley, who, it is argued, accumulated the cadavers found at Point San Jose. As such, it serves as a foundation for putting in perspective and approaching the general questions that this volume addresses.

Archaeology

The second section begins the focus on Point San Jose. It establishes the archaeological foundations of the site by addressing questions involving when the pit was used, the manner of disposal, and its meaning. It includes the pit's excavation and context, description and interpretation of artifacts, and identification and interpretation of the faunal remains. As such, this section focuses the historical background of the previous section and serves as a transition to the analysis of the Point San Jose human remains.

Human Osteology

This section continues the concentration on Point San Jose and constitutes the heart of the volume. It presents the Point San Jose human taphonomy, element counts and addresses commingling, paleodemography, stature estimations, isotopic indications of residence and diet, and paleopathology. This section assesses many of the general questions asked. In particular, it emphasizes the purpose that the remains had served, element representation, who the people were, and where they originated, as well as indicators of stress and diet.

Conclusions

Following the previous sections' concentration on Point San Jose, the final section returns to a broader perspective. It examines the Point San Jose materials and project from a sociological perspective and the volume concludes with presentation of the most salient points as well as studies yet to be undertaken. This section concludes the work and places the project in a critical theoretical framework.

References Cited

Anderson, T.

2002 A 19th Century Post-mortem Specimen from Deal, Kent. *International Journal of Osteoarchaeology* 12(3):216–219.

Angel, J. Lawrence, Jennifer O. Kelley, Michael Parrington, and Stephanie Pinter

1987 Life Stresses of the Free Black Community as Represented by the First African Baptist Church, Philadelphia, 1823–1841. *American Journal of Physical Anthropology* 74:213–229.

Black, S., G. Rutty, S. V. Hainsworth, and G. Thomson (editors)

2017 *Criminal Dismemberment: Forensic and Investigative Analysis.* CRC Press, Boca Raton, Florida.

Blakely, Robert L., and Judith M. Harrington (editors)

1997 *Bones in the Basement: Postmortem Racism in Nineteenth-Century Medical Training.* Smithsonian Institution Press, Washington, DC.

Boston, C., and H. Webb

2012 Early Medical Training and Treatment in Oxford: A Consideration of the Archaeological and Historical Evidence. In *Anatomical Dissection in Enlightenment England and Beyond: Autopsy, Pathology and Display,* edited by P. D. Mitchell, pp. 43–67. Ashgate, Farnham, United Kingdom.

Bruwelheide, Karin S., Douglas W. Owsley, Beverly A. Straube, and Jaime E. May

2017 Evidence for Early Seventeenth-Century Surgery and Dissection at Fort James, Virginia. In *The Bioarchaeology of Dissection and Autopsy in the United States,* edited by Kenneth Nystrom, pp. 41–60. Springer International, Heidelberg, Germany.

Bugaj, U., M. Novak, and M. Trzeciecki

2013 Skeletal Evidence of a Post-mortem Examination from the 18th/19th Century Radom, Central Poland. *International Journal of Paleopathology* 3(4):310–14.

Buzon, Michele R., Phillip Walker, Francine Drayer Verhagen, and Susan L. Kerr

2005 Health and Disease in Nineteenth-century San Francisco: Skeletal Evidence from a Forgotten Cemetery. *Historical Archaeology* 39(2):1–15.

Chamberlain, Andrew T.

2006 *Demography in Archaeology.* Cambridge University Press, Cambridge, United Kingdom.

Chamberlain, A.

2012 Morbid Osteology: Evidence for Autopsies, Dissection and Surgical Training from the Newcastle Infirmary Burial Ground (1753–1845). *Anatomical Dissection in Enlightenment England and Beyond: Autopsy, Pathology and Display,* edited by P. D. Mitchell, pp. 23–42. Ashgate, Farnham, United Kingdom.

Chapman, Ellen, and Mark Kostro

2017 A Dissection at the Coffeehouse? The Performance of Anatomical Expertise in Colonial America. In *The Bioarchaeology of Dissection and Autopsy in the United States,* edited by Kenneth C. Nystrom, pp. 61–76. Springer International, Heidelberg, Germany.

Charlier, P., I. Huynh-Charlier, J. Poupon, E. Lancelott, P. F. Campos, D. Favier, et al.
2013 A Glimpse into the Early Origins of Medieval Anatomy through the Oldest Conserved Human Dissection (Western Europe, 13th c. AD). *Archives of Medical Science.* doi: 10.5114/aoms.2013.33331.

Cook, Della Collins, and Mary Lucas Powell
2006 The Evolution of American Paleopathology. In *Bioarchaeology: The Contextual Analysis of Human Remains,* edited by J. E. Buikstra and L. Beck, pp. 281–323. Routledge, Abingdon, United Kingdom.

Crist, Thomas A., and M. H. Crist
2011 Skeletal Evidence for the Dissection of Children at the Philadelphia Almshouse, 1732–1834. *American Journal of Physical Anthropology* 144(S52):117.

Crist, Thomas A., Douglas B. Mooney, and Kimberly A. Morrell
2017 The Mangled Remains of What Had Been Humanity: Evidence of Autopsy and Dissection at Philadelphia's Blockley Almshouse, 1835–1895. In *The Bioarchaeology of Dissection and Autopsy in the United States,* edited by Kenneth C. Nystrom, pp. 259–278. Springer International, Heidelberg, Germany.

Crist, Thomas A., and Marcella H. Sorg
2017 Renaissance Anatomy in the Americas: A Bioarchaeological Perspective on the Earliest Skeletal Evidence of Autopsy in the New World. In *The Bioarchaeology of Dissection and Autopsy in the United States,* edited by Kenneth C. Nystrom, pp. 25–40. Springer International, Heidelberg, Germany.

Crist, Thomas A., Marcella H. Sorg, M. Larocque, and M. H. Crist
2004 *Champlain's Cemetery: Skeletal Analysis of the First Acadians, St. Croix Island International Historic Site, Calais, Maine.* Report to the US National Park Service, Acadia National Park, Bar Harbor, Maine.

Davidson, James M.
2007 "Resurrection Men" in Dallas: The Illegal Use of Black Bodies as Medical Cadavers (1900–1907). *International Journal of Historical Archaeology* 11(3):193–220.

Dittmar, J. M., and P. D. Mitchell
2015 A New Method for Identifying and Differentiating Human Dissection and Autopsy in Archaeological Human Skeletal Remains. *Journal of Archaeological Science: Reports* 3:73–79.

Dougherty, Sean P., and Norman C. Sullivan
2017 Autopsy, Dissection, and Anatomical Exploration: The Postmortem Fate of the Underclass and Institutionalized in Old Milwaukee. In *The Bioarchaeology of Dissection and Autopsy in the United States,* edited by Kenneth C. Nystrom, pp. 205–235. Springer International, Heidelberg, Germany.

Fowler, L., and N. Powers
2012 Patients, Anatomists and Resurrection Men: Archaeological Evidence for Anatomy Teaching at the London Hospital in the Early Nineteenth Century. In *Anatomical Dissection in Enlightenment England and Beyond: Autopsy, Pathol-*

ogy and Display, edited by P. D. Mitchell, pp. 77–94. Ashgate, Farnham, United Kingdom.

Goff, Alaina K.

2009 *Report of Skeletal Remains Excavated at Fort Craig Post Cemetery.* Manuscript on File with the US Department of Interior, Bureau of Reclamation, Upper Colorado Region, Albuquerque, New Mexico.

Grauer, Anne L., Vanessa Lathrop, and Taylor Timoteo

2017 Exploring Evidence of Nineteenth Century Dissection in the Dunning Poorhouse Cemetery. In *The Bioarchaeology of Dissection and Autopsy in the United States,* edited by Kenneth C. Nystrom, pp. 301–313. Springer International, Heidelberg, Germany.

Halling, Christine L., and Ryan M. Seidemann

2017 Structural Violence in New Orleans: Skeletal Evidence from Charity Hospital's Cemeteries, 1847–1929. In *The Bioarchaeology of Dissection and Autopsy in the United States,* edited by Kenneth C. Nystrom, pp. 165–184. Springer International, Heidelberg, Germany.

Harrington, Judith M., and Robert L. Blakely

1995a Bones in the Basement: Bioarchaeology of Historic Remains in Nonmortuary Contexts. In *Bodies of Evidence: Reconstructing History through Skeletal Analysis,* edited by Anne L. Grauer, pp. 105–120. Wiley-Liss, New York.

1995b Rich Man, Poor Man, Beggar Man, Thief: The Selectivity Exercised by Graverobbers at the Medical College of Georgia, 1837–1887. In *Grave Reflections: Portraying the Past through Cemetery Studies,* edited by Shelley R. Saunders and Ann Herring, pp. 153–178. Canadian Scholars Press, Toronto.

Heilen, Michael P. (editor)

2012 *Uncovering Identity in Mortuary Analysis: Community-Sensitive Methods for Identifying Group Affiliation in Historical Cemeteries.* Left Coast Press, Walnut Creek, California.

Henderson, D., M. Collard, and D. A. Johnston

1996 Archaeological Evidence for 18th-century Medical Practice in the Old Town of Edinburgh: Excavations at 13 Infirmary Street and Surgeons' Square. *Proceedings of the Society of Antiquaries of Scotland* 126:929–941.

Hillson, S., T. Waldron, B. Owen-Smith, and L. Martin

1998 Benjamin Franklin, William Hewson and the Craven Street Bones. *Archaeology International* 2:14–16.

Hodge, Christina J.

2012 Non-Bodies of Knowledge: Anatomized Remains from the Holden Chapel Collection, Harvard. *Journal of Social Archaeology* 13(1):122–149.

Hodge, Christina J., Jane Lyden Rousseau, and Michèle E. Morgan

2017 Teachings of the Dead: The Archaeology of Anatomized Remains from Holden Chapel, Harvard University. In *The Bioarchaeology of Dissection and Autopsy in the United States,* edited by Kenneth C. Nystrom, pp. 115–142. Springer International, Heidelberg, Germany.

Hull, G., P. Blinkhorn, P. Cannon, S. Hamilton-Dyer, C. Salter, and B. White

2003 The Excavation and Analysis of an 18th-century Deposit of Anatomical Remains and Chemical Apparatus from the Rear of the First Ashmolean Museum (now The Museum of the History of Science), Broad Street, Oxford. *Post-medieval Archaeology* 37(1):1–28.

Larsen, Clark Spencer

2015 *Bioarchaeology: Interpreting Behavior from the Human Skeleton.* Cambridge University Press, Cambridge, United Kingdom.

Lowe, Kim Lusignan

2017 A Historical and Osteological Analysis of Postmortem Medical Practices from the Albany County Almshouse Cemetery Skeletal Samples in Albany, New York. In *The Bioarchaeology of Dissection and Autopsy in the United States,* edited by Kenneth C. Nystrom, pp. 315–333. Springer International, Heidelberg, Germany.

Lusignan, Kimberly A.

2004 A Historical and Osteological Analysis of Postmortem Medical Practices from the Albany County Almshouse Cemetery Skeletal Sample. Master's thesis, Department of Anthropology, State University of New York at Albany.

Mann, Robert W., Douglas W. Owsley, and Paul A. Shackel

1991 A Reconstruction of 19th-Century Surgical Techniques: Bones in Dr. Thompson's Privy. *Historical Archaeology* 25:106–112.

McFarlin, Shannon C., and Lawrence E. Wineski

1997 The Cutting Edge: Experimental Anatomy and the Reconstruction of Nineteenth-Century Dissection Techniques. In *Bones in the Basement: Postmortem Racism in Nineteenth-Century Medical Training,* edited by Robert L. Blakely and Judith M. Harrington, pp. 107–161. Smithsonian Institution Press, Washington, DC.

Murphy, S. L., J. Q. Xu, K. D. Kochanek, and E. Arias

2018 Mortality in the United States, 2017. National Center Health Statistics, Data Brief, no 328. Hyattsville, MD: National Center for Health Statistics. Accessed August 28, 2019, https://www.cdc.gov/nchs/products/databriefs/db328.htm.

Novak, Shannon A.

2017 Partible Persons or Persons Apart: Postmortem Interventions at the Spring Street Presbyterian Church, Manhattan. In *The Bioarchaeology of Dissection and Autopsy in the United States,* edited by Kenneth C. Nystrom, pp. 87–111. Springer International, Heidelberg, Germany.

Novak, Shannon, and Wesley Willoughby

2010 Resurrectionists' Excursions: Evidence of Postmortem Dissection from the Spring Street Presbyterian Church. *Northeast Historical Archaeology* 39(1):134–152.

Nystrom, Kenneth C.

2011 Postmortem Examinations and the Embodiment of Inequality in 19th Century United States. *International Journal of Paleopathology* 1(3–4):164–172.

2014 The Bioarchaeology of Structural Violence and Dissection in the 19th-Century United States. *American Anthropologist* 116(4):765–779.

Nystrom, Kenneth C., Joyce Sirianni, Rosanne Higgins, Douglas Perrelli, and Jenifer L. Liber Raines

2017 Structural Inequality and Postmortem Examination at the Erie County Poorhouse. In *The Bioarchaeology of Dissection and Autopsy in the United States*, edited by Kenneth C. Nystrom, pp. 279–300. Springer International, Heidelberg, Germany.

Owsley, Douglas W.

1995 Contributions of Bioarchaeological Research to Knowledge of Nineteenth-century Surgery. In *Grave Reflections: Portraying the Past through Cemetery Studies*, edited by Shelley R. Saunders and Ann Herring, pp. 119–151. Canadian Scholars, Toronto.

Owsley, Douglas W., Karin S. Bruwelheide, Richard L. Jantz, Jodi L. Koste, and Merry Outlaw

2017 Skeletal Evidence of Anatomical and Surgical Training in Nineteenth-Century Richmond. In *The Bioarchaeology of Dissection and Autopsy in the United States*, edited by Kenneth C. Nystrom, pp. 143–164. Springer International, Heidelberg, Germany.

Owsley, Douglas W., Robert W. Mann and Sean P. Murphy

1991 Injuries, Surgical Care and Disease. In *Snake Hill: An Investigation of a Military Cemetery from the War of 1812*, edited by Susan Pfeiffer and Ronald F. Williamson, pp 198–226. Dundurn, Toronto.

Pfeiffer, Susan

1991 Estimation of Age at Death. In *Snake Hill: An Investigation of a Military Cemetery from the War of 1812*, edited by Susan Pfeiffer and Ronald F. Williamson, pp. 167–175. Dundurn, Toronto.

Richards, Patricia B., Catherine R. Jones, Emily Mueller Epstein, Nicholas W. Richards, Brooke L. Drew, and Thomas J. Zych

2017 You Couldn't Identify Your Grandmother if She Were in that Party: The Bioarchaeology of Postmortem Investigation at the Milwaukee County Poor Farm Cemetery. In *The Bioarchaeology of Dissection and Autopsy in the United States*, edited by Kenneth C. Nystrom, pp. 237–257. Springer International, Heidelberg, Germany.

Ross, Ann H., and Eugenia Cunha (editors)

2019 *Dismemberments: Perspectives in Forensic Anthropology and Legal Medicine* Academic, Elsevier, London.

Saul, Frank P., and Julie M. Saul

1986a As Though It Were a Crime: "Clues" from a Dissecting "Room" Refuse Pit. Abstracts of the Thirty-Eighth Annual Meeting of the American Academy of Forensic Sciences, New Orleans, Louisiana.

1986b Duelists, Murderers and Paupers? Osteobiography of an Early Dissecting "Room" Population. Abstracts of the Thirteenth Annual Meeting of the Paleopathology Society and poster at the 55th Annual Meeting of the American Association of Physical Anthropologists, Albuquerque, New Mexico.

1987 An Osteobiographic Analysis of an Early Dissecting "Room" Population. Abstracts of the 100th Annual Meeting of the American Association of Anatomists, Washington, DC. *The Anatomical Record* 218:1:120A.

Saunders, Shelley R.
1991 Sex Determination, Stature and Size and Shape Variation of the Limb Bones. In *Snake Hill: An Investigation of a Military Cemetery from the War of 1812,* edited by Susan Pfeiffer and Ronald F. Williamson, pp. 176–197. Dundurn, Toronto.

Stevens, William D., Carlina de la Cova, Christopher Judge, and Christopher Young
2018 Skeletal Remains from the School of Anatomy, DeSaussure College, University of South Carolina. In *Bioarchaeology of the American Southeast,* edited by S. C. Hodge and K. A. Shuler, pp. 197–224. University of Alabama Press, Tuscaloosa.

Stone, Pamela K. (editor)
2018 *Bioarchaeological Analyses and Bodies: New Ways of Knowing Anatomical and Archaeological Skeletal Collections.* Springer International, Cham, Switzerland.

Stubblefield, Phoebe R.
2011 The Anatomical Diaspora: Evidence of Early American Anatomical Traditions in North Dakota. *Journal of Forensic Sciences* 56(5):1324–1327.

Valentin, F., and F. d'Errico
1995 Skeletal Evidence of Operations on Cadavers from Sens (Yonne, France) at the End of the XVth Century. *American Journal of Physical Anthropology* 98(3):375–390.

Western, A. G.
2012 A Star of the First Magnitude: Osteological and Historical Evidence for the Challenge of Provincial Medicine at the Worcester Royal Infirmary in the Nineteenth Century. In *Anatomical Dissection in Enlightenment England and Beyond: Autopsy, Pathology and Display,* edited by P. D. Mitchell, pp. 23–41. Ashgate, Farnham, United Kingdom.

Western, A., and J. Bekvalac
2015 Digital Radiography and Historical Contextualisation of the 19th Century Modified Human Skeletal Remains from the Worcester Royal Infirmary, England. *International Journal of Paleopathology* 10:58–73.

I

Historical Context

The first of the three major sections presents historical background, including that of the San Francisco Bay Area, nineteenth-century cadaver acquisition, the Army Medical Museum and roles of post–Civil War army surgeons, and the surgeon most likely responsible for the acquisition and disposal of the Point San Jose remains. These chapters provide the historical context essential for understanding the Point San Jose remains and artifacts, and serve as the foundation for the rest of the work. They address several of this work's general themes. Those themes include what the disposal of the Point San Jose bones and artifacts represented, how the Point San Jose deposit reflected aspects of post–Civil War military medicine, and who commanded gathering, use, and disposal of the human remains. As the opening chapters of this volume, this first section lays the groundwork for understanding the archaeological and bioarchaeological interpretations presented later.

1

Historical Background of California and the San Francisco Bay Area

LISA N. BRIGHT

From the late eighteenth to early twentieth century, California, and especially the San Francisco Bay Area, underwent significant demographic, political, and cultural changes. The strategic location of the San Francisco Bay necessitated the development and installation of seacoast defenses by each reigning government. Point San Jose, a fort located on the northeastern tip of the San Francisco Peninsula, developed in response to a multitude of regional and global events. Point San Jose has seen a variety of uses from its early beginnings as a Spanish garrison to its expansion and increased artillery installation during the Civil War, to its role as a refugee camp during the 1906 earthquake. It housed both military personnel and private citizens who left their own imprint on the fort. The Bay Area, and Point San Jose specifically, has a broad and varied history that this summary only begins to explore. The pit containing human remains discovered at Fort Mason is discussed and interpreted in the following chapters, but its very presence is tied to the unique history of both the San Francisco Bay Area and Point San Jose. This historical context provides a summary of the pertinent events surrounding the European exploration and settlement of the San Francisco Bay Area, from the Spanish period through the 1906 earthquake.

San Francisco Bay Area

In geologic terms the San Francisco Bay is a relatively recent landform, reaching its present size approximately 10,000 years ago (Sloan 2006). The greater San Francisco Bay Area, popularly referred to as the Bay Area, consists of the nine counties that border the bay (Marin, Sonoma, Napa, So-

Map 1.1. California showing San Francisco Bay Area in gray. Created by Judy Stolen.

lano, Contra Costa, Alameda, Santa Clara, San Mateo, and San Francisco) (Map 1.1). The Bay Area is essentially a single large metropolis that is geographically split into several parts: North Bay (Marin, Sonoma, Napa, Solano Counties), East Bay (Contra Costa and Alameda Counties), South Bay (Santa Clara County), the Peninsula (San Mateo County), and San Francisco (San Francisco County). Today the Bay Area is home to over 7 million people (US Census n.d.). The history of the Bay Area is complex and can be divided into three periods: Spanish (1769–1821), Mexican (1821–1848), and American (1848–present).

Spanish Period: 1769–1820

The Spanish were the first Europeans to document the San Francisco Bay Area. Spanish exploration of the northern coast of Alta California began in the sixteenth century, but San Francisco Bay was not formally explored until 1769 when José de Ortega discovered the bay while scouting on top of Sweeney Ridge during Gaspar de Portolá's Pacific Coast voyage. Spanish colonial policy was directed toward establishing formal outposts in all Spanish territories. Spain's method of colonization consisted of three complementary institutions: the pueblo, the mission, and the presidio (Garcia

1997). The pueblo was the civilian village, the mission was the religious component founded by Franciscan missionaries, and the presidio was the fortified military garrison. Between 1769 and 1822 the Franciscans founded 21 missions in Alta California,[1] from San Diego in the south to San Francisco Solano (present-day Sonoma) in the north (Senkewicz 2002). The San Francisco Presidio, located on the southern side of the Golden Gate, was founded in July 1776. The Presidio was the first defensive fortification constructed and is the lone non-coastal military fortification in the San Francisco area. Mission San Francisco de Asís, generally referred to as Mission Dolores, was established west of Mission Bay. The mission, Presidio, and the small settlement of Yerba Buena landing were San Francisco's main settlements until the 1840s (Scott 1985).

Increasing concerns about potential British and Russian advancement into Alta California spurred increasing armaments at the Presidio. War with Great Britain in 1797 caused the additional strengthening of defenses throughout California. Additional batteries were erected including a small artillery installation at la Batería de San José (c. 1797), a point later developed by the Americans as Point San Jose (and became Fort Mason in 1882; Sebby 2009). This location proved to be troublesome and the battery routinely suffered extensive storm damage. By 1806 Governor Arrillaga noted that exposure had rendered the guns at these locations useless (Sebby n.d.; Thompson 1979).

Despite Spain's initial exploration and colonization of Alta California, it slowly lost both territory and power in the Americas. During the final decade of Spanish California, revolutionary movements in Spain and Mexico disrupted nearly all state support for the California outpost. Following the Mexican War of Independence (1810–1821), Alta California became a territory of Mexico.

Mexican Period: 1821–1848

The shift to Mexican rule had dramatic impacts on the missions and surrounding community. By 1835, the same year that the village of Yerba Buena (today's San Francisco) was founded, the Spanish fort Castillo de San Joaquín was abandoned, and Mexican authorities showed no interest in reviving the fort and auxiliary battery. The Presidio had also fallen into decay and was abandoned by the Mexican military (Thompson 1979).

By the 1840s American immigrants were arriving in larger numbers. Preludes to war were evident in California as early as 1842 when Mexican-appointed governor General Manuel Micheltorena's undisciplined and rowdy governance began to alienate the Californios.[2] Tensions were high, and by 1844 a small civil war began. The 1844 revolution was brief, but it created uncertainty among the foreign-born population regarding the province's power base (Garcia 1997). By 1845 several nations had challenged Mexico's claim on Alta California. Mexico's hold on the California territory was short-lived. In April of 1846 war began between the US and Mexico, stemming from the US annexation of Texas. Members of the Bear Flag Revolt decided to capture and occupy the derelict fort Castillo de San Joaquín. John Fremont claimed that this action effectively freed all of California north of the San Francisco Bay from Mexican authority (Thompson 1979). Captain John Berrien Montgomery, US Navy, also landed in Yerba Buena with a force of 70 sailors and marines and declared American possession of the land. Montgomery ordered the construction of a battery at the foot of Telegraph Hill. The battery was aptly named Fort Montgomery, although that name was considered unofficial. The battery was armed with guns from the defunct Castillo de San Joaquín. For the next three years, until the Treaty of Guadalupe Hidalgo was signed, California was ruled by American military governors, all the while still technically the territory of Mexico.

US Navy forces and the California Battalion, a volunteer militia it authorized, captured Monterey and San Francisco, eventually occupying every presidio and pueblo. American forces overpowered the opposing Californio forces (Haas 1998), although some did not oppose the impending American takeover and laid down their arms or joined them (Pfaelzer 2007). They were instructed to adhere to Mexican law when it did not conflict with American objectives, but rampant land speculation and continuous immigration into the territory hinted at the momentous changes that were to come (Haas 1998). In early 1848 Colonel Richard Mason ordered a reserve be set aside at the Presidio and Point San Jose. On February 2, 1848, Mexico signed the Treaty of Guadalupe Hidalgo and ceded Alta California to the US. The boundaries established by Mason's order were formalized by President Fillmore after California was ceded to the US (Pfaelzer 2007; Thompson 1979).

American Period: 1848–Present

Gold Rush Period: 1849–1860

Although the influx of immigrants before Alta California was ceded seemed daunting, it was nothing compared to what was to come. On January 24, 1848, just nine days before the signing of the Treaty of Guadalupe Hidalgo, gold was discovered in the Sierra foothills at Sutter's Mill on the American River (Beck and Haase 1974; Garcia 1997; Pitti 2003). The Gold Rush changed California and precipitated a population boom, one of the largest mass migrations in US history. That influx created a soaring demand for goods, land, and housing. These products had little or no direct connection to gold mining and fueled an economic expansion that benefited other industries (St. Clair 1998). The discovery of gold in the California foothills turned San Francisco from a quiet port town into a hub of commerce and immigration.

By 1850 nearly every canyon and small valley in central and northern California had a small town or miners' camp. San Francisco had been transformed from a small village into a major city of over 50,000 and established its position as the commercial, financial, and shipping center of California (Scott 1985). The Gold Rush created a quickly changing landscape, providing the impetus toward the development of roads, the growth of freight companies, and the emergence of a stagecoach system. The immense size of California coupled with its remoteness from other major US population centers combined to make railroads vital to the state's development. Previously the principal route to California had been by water, but the completion of the transcontinental railroad in 1869, as well as other local railroad lines throughout the state, made migration and commerce much easier (Beck and Haase 1974).

The increasing importance of the San Francisco Harbor, and the protection of the vast amounts of gold passing through the Golden Gate, heightened the US government's need to protect the bay and strained the existing military infrastructure. The government needed to guarantee the safety of ships entering and exiting the harbor. Thus, in 1851 President Fillmore formally established the Point San Jose Military Reservation, separate from the Presidio. The US Army began building extensive masonry fortifications at Fort Point and on Alcatraz Island (Thompson 1979). Two batteries were planned for Point San Jose. Although it had been set aside as a military reserve in 1850, the government had not developed the property.

Ambitious affluent individuals quietly moved onto the military land and began constructing homes (National Park Service [NPS] 2015a, 2015b, 2019; Thompson 1979).

Civil War Period: 1861–1865

The outbreak of the American Civil War had immediate repercussions on California. Because the war cut off the Union from many of its resources, the San Francisco harbor became increasingly important for its proximity to the wealth emerging from the gold and precious mineral mines. Although there was a substantial amount of pro-Union sentiment in California, there was also a very vocal pro-Confederacy faction. Despite the admission of California to the Union as a free state, several hundred slaves worked in the mines and as domestic servants. The Fugitive Slave Act of 1852 provided that state officers must assist in the return of fugitive slaves, but conversely also suggested that slaves who were brought in California before its 1850 statehood may have protected status and could not be claimed as property and returned to their former owners. The ambiguity of the act led to many disputes regarding the legal rights of both fugitive slaves and slave owners in the state (Matthews 2012; Richards 2009).

Although physically distant from the principal theaters of war, California made efforts to defend itself once war broke out and determined how it might contribute to the war effort. In 1849 a joint Army-Navy commission had met in San Francisco and determined it strategically advantageous to construct two forts across from one another at the narrowest point of the bay entrance. These forts would also be supplemented by fortifications and gun batteries on Alcatraz Island to repel foreign invaders. In the 1860s fear of Confederate incursion fueled further fortification efforts (Matthews 2012; Richards 2009).

As previously stated, the military threat to California from the Confederacy was a major concern for Unionists during the Civil War. San Francisco Harbor was particularly vulnerable to attack by sea. To meet this potential threat, heavy smoothbores and rifled guns were introduced and the expansion of fortifications, including Angel Island, Alcatraz Island, and Point San Jose, commenced. Point San Jose was known to locals as "Black Point" and had become home to a group of wealthy citizens, including John Fremont of the Bear Flag Revolt. Although the squatters protested the development plans, the outbreak of the Civil War forced the army to evict the civilian residents and retake possession of Point San Jose. The army constructed two batteries at the northern tip of the point. The military expansion consisted of newly constructed headquarters, hospital, barracks, stables, and outbuildings centered around a parade ground. Not all the civilian houses were demolished; some were converted into officer housing (NPS 2015a, 2015b, 2019; Thompson 1979). The Point San Jose skeletal assemblage was recovered adjacent to the hospital building. The military construction history, site stratigraphy, and artifact analysis concluded that the bone pit originated from a single depositional event sometime between 1864–1891 (see Locke Barton, and Gavette and Willey, this volume).

The navy ordered the iron-clad warship *Camanche* to patrol the bay. The Confederate raider *Shenandoah* planned to breach the bay on July 5, 1865, only to discover that the war had ended, and the Confederacy had been defeated. The *Shenandoah* never made it to San Francisco Bay and continued their raiding in the north Pacific before heading to England. Thus, no enemy entered the bay during this period and no shot was fired from the newly installed guns around the harbor entrance (Gilbert 1954, 1965; Miller 1966; NPS 2015a, 2015b, 2019; Thompson 1979).

Post–Civil War Period: 1866–1906

The period following the Civil War focused on renovating existing batteries and military installations based on lessons learned during the Civil War, which made masonry works obsolete. Technological updates were planned for Fort Point, Lime Point, Point San Jose, Alcatraz Island, and Angel Island. Although recommendations were made for Point San Jose to undergo a significant expansion, because the fort was viewed as secondary to Lime Point and Fort Point, the construction was not a high priority. The temporary batteries constructed during the 1860s continued to be the only points of defense for the next 30 years. In 1882 Point San Jose was renamed Fort Mason in honor of Colonel Richard Barnes Mason, the second governor and commander of California (NPS 2019; Thompson 1979).

As the US expanded its military power, the Army was tasked with increasing the defenses of the nation's coasts and ports. In 1885 President Cleveland established the Endicott Board for the purpose of modernizing military fortifications. The board deemed San Francisco harbor the most vital strategic location on the West Coast. New forts, batteries, and guns were proposed for the harbor entrance, while the existing fortifications at Fort Baker, Fort Barry, and Fort McDowell saw their defensive capabilities increased. Fort Mason changed very little. An emergency battery was constructed to defend against a potential attack during the Spanish-American

War (1898), and a single additional battery was built in 1899. The battery was disarmed in 1909 (Sebby n.d.; Thompson 1979).

After the Civil War, the Bay Area continued to see exponential economic growth. However, as the rest of the California, the region suffered from a lack of rapid overland communication with the eastern US and from a lack of railroad transportation for both goods and people. The smaller regional rail lines and completion of the transcontinental railroad in 1869 helped spur the Bay Area's economic growth. The area's wealth transitioned from the gold mines to agricultural production, factories, banking, and the iron and steel industry.

By the late 1860s, San Francisco dominated an economic empire that stretched from the Pacific Coast to the Rockies. San Francisco in the 1880s and early 1890s was on the way to becoming one of the most densely populated cities in the US. By 1900 it enjoyed one of the greatest economic booms in its history with a steady influx of immigrants and migrants to the city pushing demand for development of new office buildings, hotels, stores, and residential buildings (Scott 1985).

1906 Earthquake

At 5:13 a.m. on April 18, 1906, a seismic event on the San Andreas Fault shook the entire Bay Area for 45–60 seconds. It was felt from southern Oregon, south to Los Angeles, and inland as far as central Nevada, an area of approximately 200,000 square miles, and estimated to register a magnitude of 8.25 on the Richter scale. Although the earthquake caused severe property damage and loss of life, the subsequent fire proved far more destructive. Fires from gas jets and chimneys broke out across the city. The dome of the San Francisco city hall collapsed, as did the dome of the Sonoma County Court House some 50 miles away in Santa Rosa. At Point Reyes Station in Marin County, a train that was about to depart to San Francisco toppled over. Every building at the Agnew State Insane Asylum in San Jose collapsed, killing more than a hundred patients and 11 staff members. At Stanford University, the chapel spire dropped through the roof. In Oakland, Alameda, Berkeley, Richmond, Vallejo, Petaluma, and San Rafael, chimneys collapsed, and while buildings were damaged as a whole, the impact was minimal. Almost all towns on the San Francisco Peninsula, however, suffered considerable damage but remained functional (Dicker 1980; Fradkin 2005).

The fire in San Francisco burned for three days. Because the principal

Figure 1.1. San Francisco earthquake, 1906: Fort Mason, where thousands of destitute sufferers found relief. Photographer H. C. White Co.

water mains in the city, as well as the pipelines that conveyed water to San Francisco from surrounding springs and reservoirs, had bent and burst during the earthquake, the fire department was helpless to combat the blaze. Where some 28,000 buildings had stood, there was nothing but destruction and ruin. The earthquake and fire destroyed over 500 city blocks, or roughly four-square miles. Estimations of casualties from the earthquake and fire range from 300–700, but they greatly undercount the causalities because there was an undetermined number of individuals buried in the rubble of collapsed buildings and killed during the fire. Additionally, it was difficult to determine who among the individuals reported missing were dead or alive when San Francisco's population scattered into refugee camps throughout the Bay Area (Dicker 1980; Fradkin 2005).

In San Francisco, 21 official refugee camps for the hundreds of thousands of displaced individuals were established in city parks and squares including Fort Mason (Figure 1.1), the Presidio, Golden Gate Park Speedway, Washington Square, and Camp Ingleside. Cities throughout the Bay Area sheltered refugees in private homes and in tent cities hastily erected by municipal authorities. The impact of the disaster rippled across the county. San Francisco was the West Coast commercial hub, so commerce was impacted throughout the West and beyond (Fradkin 2005). The Bay Area communities quickly rebuilt following the earthquake, and the Bay Area maintained its dominance as a manufacturing, transit, agricultural, and commercial hub (Scott 1985).

Conclusion

Point San Jose grew slowly but was drastically altered each time possession of California changed hands. Although it began as a small, poorly maintained Spanish artillery installation, American military expansion precipitated by the Civil War greatly expanded the number of individuals stationed at the point, as well as the number of buildings and fortifications. While this chapter is but a brief overview of the 150 years of global, national, and regional history associated with Point San Jose, all of these contexts coalesced to create the necessary setting for the formation of the pit containing bones, buttons, and bottles that was uncovered in 2010 adjacent to the hospital building.

The following chapters in this section of the volume will speak in more detail about the background associated with the study of anatomy and cadaver treatment (Milligan, this volume), the role of US Army surgeons (Spatola and Pearlstein, this volume), and a discussion of Dr. Edwin Bentley, Point San Jose post surgeon from 1869–1876, and the most likely individual responsible for the anatomization of the human remains recovered in the Point San Jose assemblage (Gavette, this volume).

Notes

1. A historical geographic region that roughly corresponds to present-day California, Nevada, Utah, and Arizona.

2. Spanish-speaking individuals living in Alta California.

References Cited

Beck, Warren, and Ynez Haase
1974 *Historical Atlas of California.* University of Oklahoma Press, Norman.

Dicker, L.M.
1980 The San Francisco Earthquake and Fire: Photographs and Manuscripts from the California Historical Society Library. *California History* 59(1):34–65.

Fradkin, P.L.
2005 *The Great Earthquake and Firestorms of 1906: How San Francisco nearly Destroyed Itself.* University of California Press, Berkley.

Garcia, Lorie
1997 *Santa Clara: From Mission to Municipality.* Santa Clara University, Santa Clara.

Gilbert, Benjamin Franklin
1954 San Francisco Harbor Defense during the Civil War. *California Historical Society Quarterly* 33(3):229–240.

1965 The Confederate Raider Shenandoah: The Exclusive Destroyer in the Pacific and the Arctic. *Journal of the West* 4(2):169–182.

Haas, Lisbeth
1998 War in California, 1846–1848. In *Contested Eden: California Before the Gold Rush*, edited by Ramón Gutiérrez and Richard Orsi, pp. 331–356. University of California Press, Berkeley.

Matthews, Glenna
2012 *The Golden State in the Civil War: Thomas Starr King, the Republican Party, and the Birth of Modern California*. Cambridge University Press, Cambridge.

Miller, Robert Ryal
1966 The *Camanche:* First Monitor of the Pacific. California Historical Society Quarterly 45(2):113–124.

National Park Service
2015a Civil War at Fort Mason. Electronic document, https://www.nps.gov/goga/learn/historyculture/civil-war-at-fort-mason.htm, accessed March 1, 2020.
2015b Harbor Defenses. Electronic document, https://www.nps.gov/goga/learn/historyculture/harbor-defenses.htm, accessed March 1, 2020.
2019 Fort Mason. Electronic document, https://www.nps.gov/goga/learn/historyculture/fort-mason.htm, accessed January 15, 2020.

Pfaelzer, Jean
2007 *Driven Out: The Forgotten War against Chinese Americans*. Random House, New York.

Pitti, Stephen
2003 *The Devil in Silicon Valley: Northern California, Race, and Mexican Americans*. Princeton University Press, Princeton.

Richards, Leonard
2009 *The California Gold Rush and the Coming of the Civil War.* Vintage Books, New York.

Scott, Mel
1985 *The San Francisco Bay Area A Metropolis in Perspective*. 2nd ed. University of California Press, Berkeley.

Sebby, Dan
2009 History of Fort Mason. Electronic document, http://www.militarymuseum.org/Pt%20San%20Jose.html, accessed December 17, 2019.

Senkewicz, Robert
2002 The California Context. In *Telling the Santa Clara Story, Sesquicentennial Voices*, edited by Russell Skowronek, pp. 20–27. McNaughton and Gunn, Saline, Michigan.

Sloan, Doris
2006 *Geology of the San Francisco Bay Region*. University of California Press, Berkley.

St. Clair, David
1998 The Gold Rush and the Beginnings of California Industry. *California History* 77(4):185–208.

Thompson, Erwin
1979 *Historic Resource Study: Seacoast Fortifications, San Francisco Harbor, Golden*

Gate National Recreation Area, California. Denver Service Center, Historic Preservation Team, National Park Service, United States Department of the Interior.

US Census

n.d. Bay Area Census. Electronic document, http://www.bayareacensus.ca.gov, accessed January 15, 2020.

US Geological Survey (USGS)

n.d. The San Andreas and Other Bay Area Faults, Electronic document, https://earthquake.usgs.gov/earthquakes/events/1906calif/virtualtour/bayarea.php, accessed January 15, 2020.

2

Resurrectionists, Criminals, and the Unclaimed

History of Cadavers and the Study of Anatomy in the Nineteenth Century

COLLEEN F. MILLIGAN

The social history of cadavers in America is intimately tied to the public welfare movement, first established in Europe and subsequently brought to America in the seventeenth century. Early procurement and use of cadavers for medical study was directly linked to social welfare policies regarding marginalized and minority members of society. Voluntary donations to science, such as present-day body bequeathal programs, were not implemented until the mid-twentieth century. For that reason, the history of cadaver use is also associated with social theory involving embodiment, agency, identity, racism, and bias (see Hall, this volume). In particular, loss of identity in life and death was a substantial concern for paupers because of the connection between the public welfare system and the treatment of those who died in the system. Paupers, indigents, poor, impoverished, vagrants, infirm, insane, dependents, and unfortunates—these are all labels given to the inhabitants of institutions of public welfare in the nineteenth century, where many of the bodies used for medical study were procured.

The recent edited volume by Kenneth Nystrom (2017a) highlights historical and bioarchaeological studies conducted on skeletal samples from institutional settings. The study of individuals from potter's fields provides evidence of the practice of autopsy and dissection in the US that existed from the seventeenth century through the twentieth century. The context of the cemeteries and pits where these skeletal remains are found shows the extent that marginalization practices persisted. Often these locations are discovered, or more appropriately rediscovered, during construction activities in areas where forgotten features occur. This rediscovery creates a multilayered loss of identity for individuals interred in public cemeteries. For most individuals, they are marginalized members of society in life, they are nameless in death, and collectively they are forgotten by future generations in their final place of rest. This chapter outlines the historical context of cadaver use and procurement as a foundation to later discussions regarding the theoretical constructs of identity for those used as cadavers (Hall, this volume). It follows the history of anatomy and dissection globally, nationally, and regionally as it pertains to the context of the Point San Jose sample.

History of Anatomy and Dissection outside the US

The Edwin Smith Papyrus is one of the oldest written documents referring to the practice of anatomy and dissection. Dating to 1700 BC, it outlines early Egyptian knowledge of anatomy (Tubbs et al. 2019). Egyptians have long been known for their practice of mummification. Because removing internal organs was one of the first steps in mummification, this practice meant that Egyptian embalmers needed anatomical knowledge to perform their duties. Even before written evidence, Egyptians were using their knowledge of human anatomy to dry and preserve the bodies (Cosans 1997; Singer 1957).

Early anatomical study varied widely between groups around the world (Tubbs et al. 2019). Some studies of anatomy relied solely on external examinations. Some anatomists used animal surrogates for understanding human anatomy. Very few groups, however, found the practice of human dissection acceptable or legal (Tubbs et al. 2019).

Formal dissection traces back to ancient Greece with the establishment of the school of Greek medicine in Alexandria during the third century BC. The empiricists of Alexandria felt that practical experiences were needed to build theoretical concepts of the function of the human body (Cosans 1997). It was here that the first ancient Greek physicians began to perform systematic dissections of human cadavers (Cosans 1997; Singer 1957; Tubbs et al. 2019). Before Alexandria, superficial study of pathological conditions served as the extent of anatomy study. Religious and moral codes often prohibited more invasive studies.

The third century CE also witnessed the rise of rationalists, who claimed that anatomical dissection held little academic value but instead suggested that the human body and its functions could be deduced through reasoning before experimental study (Cosans 1997). After the loss of Alexandria in AD 389, anatomical studies remained largely stagnant. The exception during this period were Arab societies, which contributed to anatomical study by preserving and teaching anatomical principles established during the earlier Greek and Roman periods (Lassak 1958). Most other regions did not develop formal anatomical studies until the twelfth-century establishment of universities in France, Italy, and Germany (Ghosh 2015; Lassak 1958; Singer 1957).

By the fifteenth century, increase in the practice of dissection began to lead to a shortage of cadavers (Ghosh 2015; Lassek 1958). The growing acceptance of the use of cadavers for medical study by physicians and surgeons meant that additional sources of cadavers were needed. Practices, such as grave robbing, became increasingly common as a result (Ghosh 2015; Magee 2001; Ritchie 1912). In England, pressure from anatomists resulted in the Murder Act of 1752, which legalized the dissection of the bodies of executed murderers (Ghosh 2015; Magee 2001). The Murder Act intended to prevent the crime of murder by the threat of being dissected after hanging, and it also created an avenue for procurement of fresh human cadavers for anatomical studies (Ghosh 2015; Lassek 1958). However, by the middle of the eighteenth century, it became evident that bodies supplied from executions were insufficient to meet the demands of anatomical study. Accordingly, many European countries passed legislations allowing the use of the unclaimed bodies and individuals who died in public welfare institutions (Ghosh 2015).

History of Anatomy Laws in the US

Similar to European medicine, anatomical studies have been the foundation of the US medical practice since the seventeenth century. The establishment of an anatomy program at the University of Pennsylvania in 1745 was the first formal instruction in the US (Tward and Patterson 2002). However, cadavers for dissection and anatomical study were scarce. Most states that set up policies for legal use of cadavers mandated the use of corpses from criminal execution for capital crimes. As in Europe, dissection was considered an additional punishment for prisoners' crimes. Throughout the seventeenth and eighteenth century, crimes that were considered taboo or particularly offensive, such as murder, robbery, arson, and theft, had the added punitive measure of having one's body sent for dissection (Bennett

2018). An example of the link between criminality and the loss of bodily rights in death was a 1784 Massachusetts law that those who died through dueling would also be potentially given to the study of dissection (Tward and Patterson 2002). The practice of dissection maintained a stigma with the general public and the use of the bodies of felons was considered one of the few avenues for study (Humphrey 1973). Early acts such as the 1784 Massachusetts law on dueling were not enacted to legalize dissection in the US. Instead, they were meant to emphasize the criminality of the illegal acts themselves (Shultz 1992).

In all states, the demand for cadavers for dissection outpaced the legal supply. In an effort to alleviate the shortage of bodies during the last three centuries, the practice of acquiring bodies for medical study included marginalized and minority groups in the US. American Blacks and paupers of all ancestries were particularly vulnerable, including recent immigrants (Davidson 2007; Nystrom 2017b; Savitt 1982). If the need for more cadavers exceeded what could be procured legally, then additional bodies were often stolen from cemeteries. The body snatchers were referred to as "resurrectionists." While this practice was illegal in all states, targeting marginalized groups provided the safest avenue to avoid prosecution (Humphrey 1973). The 1788 Resurrection Riot in New York showed that when more prominent members of society were used in anatomy laboratories, the potential for public demonstrations and even violence was increased (Hulkower 2011). This method of body procurement for anatomical study accounted for thousands of cadavers annually during the eighteenth and early nineteenth centuries (Richardson 1988).

Like criminals, individuals who died in almshouses, asylums, and indigent hospitals were also seen as individuals who forfeited the right to their body in death. The movement toward using the bodies of the unclaimed from social welfare institutions was not just practiced in the US. It followed a move involving institutionalization of the poor and sick worldwide during the nineteenth and twentieth centuries. As an example, Richardson (1988) writes that between 1832 and 1932, 57,000 bodies were dissected in London's anatomy programs. The bodies reportedly used came primarily from pauper institutions, with less than half of a percent coming from other sources (Richardson 1988).

To understand the use of unclaimed bodies from social welfare institutions, there is also a need to understand the organization of these institutions themselves. In 1905 Charles Lawrence published a history of the Philadelphia Almshouse covering its 200 years of existence. The volume provides insight into the early development of indigent care in the US because the struggles of the Philadelphia Almshouse reflect nationwide trends. Institutions such as almshouses encompassed both short-term and long-term residential stays. They were often associated with indigent hospitals, workhouses, asylums, and dependent children's homes. The nature of all of these institutions also necessitated nearby, accessible cemeteries. Lawrence provides detailed notes from an early board of directors' meeting on March 3, 1790 regarding the use of the municipal potter's field. This discussion revolved around continued use of the burial lot and the utility of using inmates in the labor " . . . and graves dug by paupers in this institution, whose labor in any other way, would, from their unskillfulness and incapacity, be very little productive for the public interest, and should the Guardians of the Poor be deprived of their present economical means of making use of the their labor in the interment of those for whose funeral expenses the public must pay" (Lawrence 1905:38).

He recounted another exchange of the board of guardians on November 10, 1845, in which several members of the board protested the procurement of bodies from the potter's field for dissection and study. The inmates informed the board of their fear of being used for dissection if they died in the almshouse. Lawrence wrote that the overall response of the board was that "the colleges must have subjects" (Lawrence 1905:161). Humphrey (1973) states that a 1913 survey found that the majority of the bodies used by the medical schools came from institutions like almshouses. Highlighting the connection between pauperism and cadaver acquisition, Lawrence (1905:160–162) further quotes the board of directors' meeting notes:

> Paupers are of no use to society while living, there is no wrong done in making them useful when dead. . . . Few, and perhaps none, are so deadened in feelings as not to desire the rites of Christian burial, for who would not revolt at the idea, if they were consulted on the subject, of permitting their bodies to be exposed in the lecture rooms, cut to pieces for the benefit of the schools and then thrown into a pit containing the remains of hundreds of others . . . if the supply from the Almshouse was cut off, the bodies would be stolen from the cemeteries; and that it was better that those who died without friends or relatives to mourn for them should go to the dissecting rooms. . . .

The development of institutionalized care for the poor and infirmed in the US began as early as the seventeenth century, but larger state- or municipal-run facilities became more common in the mid-nineteenth century.

Many of these institutions were managed by a board of directors with physicians overseeing the general care of the inmates. The institutions provided places of care for those persons unable to care for themselves (Katz 1986). Their disabilities ranged from severe medical or mental health issues to individuals either without families or whose families could not support them. Inmate populations also disproportionately included recent immigrants to the US (Humphrey 1973). The cemeteries associated with the institutions were known as "potter's fields," indigent grounds, or public cemeteries. Often the physicians or the institutions were associated with medical colleges. The already loose connections to the larger community and the proximity of medical institutions made the use of the bodies of inmates from these institutions common.

At the height of grave robbing during the nineteenth century, New York and Massachusetts passed anatomy acts that legalized the dissection of unclaimed bodies (Humphrey 1973; Hulkower 2011). Individuals or families unable to pay burial fees were often included in this category. More states followed suit after the Civil War (Humphrey 1973). The tie between pauperism or marginalization in American society and cadaver study persisted through the middle of the twentieth century. Public sentiment toward dissection was primarily negative and the act of donating one's body for study was not yet a widespread practice.

Even while efforts were being made to change societal attitudes toward dissection and donation in the middle of the twentieth century, bequeathal remained a little-employed alternative to traditional burial. Over the years, numerous divergent anatomy acts or laws made body donation legally complicated (Hulkower 2011). Noting this legal difficulty and increased interest in body donation, the National Conference of the Commissioners on Uniform State Laws approved the Uniform Anatomical Gift Act (UAGA) in 1968 (Ghosh 2015). UAGA officially made body donation a right, based on a voluntary free choice (Jones 1994). Today every state has some form of law regarding body donation and donor's rights (Ghosh 2015). However, even with these advances, the legal rights of a person over his or her own body after death remain ambiguous. While body donations comprise a significant portion of all cadavers used in medical study, use of unclaimed bodies is still practiced throughout the US (Humphrey 1973; Hulkower 2011).

History of California Anatomy Laws

The history of cadaver acquisition and use in California followed a similar pattern as the rest of the US. When California became a state, many other states had already established anatomy laws. As a result, California based its early anatomy laws on standards created by those states. The first of these laws, passed in 1854, strictly forbid the practice of dissection. It further penalized any disturbance of public cemeteries (Meyer 1930). However, as medical education developed in California, the need for materials for anatomy and dissection showed a similar increase as elsewhere in the US.

One early supporter of the organization of medical societies in California was Dr. Elias Cooper. He settled in California from Peoria, Illinois, in 1855. Cooper was a surgeon by training whose medical practice in the Midwest included a dissection room for study. He taught private anatomy classes when cadavers from executed criminals were available (Wilson 1998). While Cooper's dissection practices seem by the available evidence to follow the law, he, as other anatomists at the time, retained anatomical specimen from the dissections and autopsies. Historical records at Stanford's Lane Medical Library suggest that Cooper also clandestinely removed material from his dissection laboratory to avoid public outcry (Wilson 1998). Whether the practices were legal or not, anatomical study carried a stigma with it as far as the general public was concerned. However, increasingly for medical studies, dissection became a necessity for adequate medical preparation as noted earlier (Ponce 2012).

Before arriving in California, Cooper co-founded the Illinois State Medical Society, opened county hospitals, and served as anatomy faculty at early Illinois medical colleges (Jones 1964; Wilson 1998). After arriving in California, he continued his involvement in medical education and organized the first medical colleges in the state. The Cooper Medical College at the University of the Pacific opened in 1858 and the Toland Medical College followed at the University of California, San Francisco, in 1863 (Meyer 1930; Wilson 1998). Both early medical colleges centered medical education around the Bay Area. Those first medical colleges in California included lectures on anatomy, although the practice of dissection was still illegal when they opened. Dissection illegality changed quickly.

In 1864, the California Legislature approved an act "to promote the study of Anatomy" (California State Legislature [CSL] 1864:321). The Act allowed any duly qualified physician or surgeon the right to possess "human dead bodies, or parts thereof, for the purposes of anatomical inquiry or instruction" (CSL 1864:321). Section 2 authorized officials, such as a coroner or warden, to release bodies for this use. Specifically, they could "surrender the dead bodies of such persons as are required to be buried at the public expense . . . to be by him used for the advancement of anatomical science" (CSL 1864:321). Those individuals who could not afford to pay for their own burial or lacked willing friends or relatives were potentially subject to medical study. Further, it stated that preference be given to "Medical Schools by law established in this State, for their use in the instruction of medical students." Another provision of the act required that the physician receiving such cadavers give a bond with two sureties, "that each body so by him received shall be used only for the promotion of anatomical Science, and that after having been so used the remains thereof shall be properly buried in some public cemetery, and the usual record made of name and burial, at the expense of the Physician being in possession of said body" (CSL 1864:321).

The Anatomy Act was amended in 1870 to change the wording of possession from "may have in his possession human dead bodies" to "may obtain, as hereinafter provided, and have in his possession human dead bodies" (Parker 1871). The amended Act changed the wording for the state authorities from "may surrender the dead bodies" to "shall surrender" as well as removing a provision for the use of "those persons who have died during their term of service in the State Prison, or been executed for crime" (Parker 1871).

This practice followed legalization of use of "unclaimed" bodies in other regions of the US. Also similar to other anatomy laws across the country, it specified that the bodies of the unclaimed could come from sheriff's offices, coroner's offices, county poorhouses, public hospitals, county jails or prisons, or San Francisco city officials (Hittel 1870; Meyer 1930). After the body was used for instruction, the physician was required to bury it at his own expense. The 1870 Anatomy Law made exemptions for the bodies of travelers, those who left specific requests to be buried, or were claimed by friends. Further exemptions were made in 1889 for ex-Union military personnel and in 1901 for all honorably discharged military personnel and their widows (Meyer 1930). As occurred in the earlier history of cadaver use, this law balanced the desires of the medical colleges with the public distaste for dissection by offering the marginalized members of society to fill the need of the colleges.

The anatomy laws of California remained consistent with the 1864 Act and the 1870 amendment until the establishment of California's Health and Safety Code in April of 1939 (California Health and Safety Code § 7200–7208). California Health and Safety Code Division 7, Part 1, Chapter 4 details the disposal of unclaimed dead in Sections 7200–7208. This part of the Health and Safety Code, first signed into law in 1939, outlines the procedures for documenting and ultimately disposing of unclaimed bodies. It states that, "[t]he bodies of the unclaimed dead retained by the State department shall be used solely for the purpose of instruction and study in the promotion of medical, chiropractic, and embalming education and science within the State." (California Health and Safety Code § 7203).

Conclusion

The Point San Jose series seems to indicate remains that were used for anatomical dissection and also possibly as part of a pathological collection. This hypothesis is supported by the osteological findings, including the biased recovery rates suggesting differential disposal (Cox et al., this volume), unexpected population demographics not representative of a military installation (Sgheiza and Willey, this volume), evidence of nutritional or disease-related stress (Milligan et al., this volume), and a high incidence of incised and saw cut marks in regions rarely explored during autopsy (Peters et al., this volume).

The context surrounding the disposition of remains at Point San Jose would have occurred during the early formation of California's Anatomy Laws and the growth of colleges of medicine in San Francisco. The surgeons assigned to the military post of Point San Jose in the Bay Area during the mid–late nineteenth century, as will be shown in later chapters, had ties to the larger medical community growing in the San Francisco area (Gavette, this volume). That included ties to San Francisco's medical colleges. As an example, Dr. Edwin Bentley, a surgeon stationed at Point San Jose starting in 1869, was also part of the anatomy faculty at the University of the Pacific Medical College for five years (Cobb 1980). The surgeons also sent specimens from Point San Jose to the newly established anatomical military museum, the Army Medical Museum in Washington DC (Spatola and Pearlstein, this volume; Leikind 1954).

As has been discussed here, early procurement of cadavers for dissection proved difficult and legally acquired cadavers often came from criminal execution, minorities, or marginalized individuals until the mid-twentieth century. When the demand for cadavers outpaced the legal supply, grave robbing by "resurrectionists" arose as a widespread and profitable business.

Whether targeted by resurrectionists or considered "unclaimed" under anatomy laws, the individuals most often targeted were minority groups. However, what all groups had in common was that the practice targeted members of lower socioeconomic status, regardless of ancestry (Davidson 2007; Nystrom 2017b; Savitt 1982). Tracing the origin of the Point San Jose remains includes both an understanding of the identifying information that can be gleaned from the remains themselves, and an understanding of the social context in which the remains may have been procured in the first place.

References Cited

Bennett, R. E.

2018 *Capital Punishment and the Criminal Corpse in Scotland, 1740–1834.* Palgrave Macmillan, London.

California State Legislature

1864 The Statutes of California, passed at the Fifteenth Session of the Legislature, 1863–4. O. M. Clayes, State Printer, Sacramento. https://leginfo.legislature.ca.gov/faces/codes_displayText.xhtml?lawCode=HSC&division=7.&title=&part=1.&chapter=4.&article=, accessed April 12, 2020.

Cobb, W. Montague

1980 Edwin Bentley, MD, 1824—1917: Connecticut Yankee in Washington and Little Rock. *Journal of the National Medical Association* 72(10):1012–1014.

Cosans, C. E.

1997 Galen's Critique of Rationalist and Empiricist Anatomy. *Journal of the History of Biology* 30(1):35–54.

Davidson, J. M.

2007 "Resurrection Men" in Dallas: The Illegal Use of Black Bodies as Medical Cadavers (1900–1907). *International Journal of Historical Archaeology* 11:193–220.

Ghosh, S. K.

2015 Human Cadaveric Dissection: A Historical Account from Ancient Greece to the Modern Era. *Anatomy & Cell Biology* 48(3):153–169.

Hittel, T. H.

1870 *The General Laws of the State of California, from 1850 to 1864, Inclusive: Being a Compilation of All Acts of a General Nature Now in Force, with Full References to Repealed Acts, Special and Local Legislation, and Statutory Constructions of the Supreme Court: to which are Prefixed the Declaration of Independence, Constitution of the United States, Treaty of Guadalupe Hidalgo, Proclamations to the People of California, Constitution of the State of California, Act of Admission and United States Naturalization Laws: with Notes of California Decisions Thereon. Volume 2.* H. H. Bancroft, Sacramento, California.

Humphrey, David C.

1973 Dissection and Discrimination: The Social Origins of Cadavers in American, 1760–1915. *Bulletin of the New York Academy of Medicine* 49:819–825.

Jones, D. G.
1994 Use of Bequeathed and Unclaimed Bodies in the Dissecting Room. *Clinical Anatomy* 7(2): 102–107.
Jones, J. R.
1964 *History of the Medical Society of the State of California.* The Historical Commission of the Sacramento Society for Medical Improvement, Sacramento, California.
Katz, M. B.
1986 *In the Shadow of the Poorhouse: A Social History of Welfare in America.* Basic Books, New York.
Lassak, A. M.
1958 *Human Dissection: It's* [*sic*] *Drama and Struggle.* Thomas, Springfield, Illinois.
Lawrence, C.
1905 *History of Philadelphia Almshouses and Hospitals.* Arno Press, New York.
Leikind, Morris C.
1954 Army Medical Museum and Armed Forces Institute of Pathology in Historical Perspective. *The Scientific Monthly* 79(2): 71–78.
Magee, R.
2001 Art Macabre: Resurrectionists and Anatomists. *ANZ Journal of Surgery* 71:377–380.
Meyer, A. W.
1930 Anatomy Acts Of California: A Survey of Former and Present Laws. *California and Western Medicine* 33(4):703–707.
Nystrom, K. C.
2017a *The Bioarchaeology of Dissection and Autopsy in the United States.* Edited by Kenneth C. Nystrom. Springer International, Switzerland.
2017b Introduction. In *The Bioarchaeology of Dissection and Autopsy in the United States,* edited by K. C. Nystrom, pp. 1–22. Springer International, Switzerland.
Parker, Charles H.
1871 *The General Laws of the State of California, from 1864 to 1871.* Vol. 3. A. L. Bancroft, San Francisco.
Ponce, R. N.
2012 "They Increase in Beauty and Elegance": Transforming Cadavers and the Epistemology of Dissection in Early Nineteenth-Century American Medical Education. *Journal of the History of Medicine and Allied Sciences* 68(3):331–376.
Richardson, R.
1988 *Death, Dissection, and the Destitute.* Penguin Books, London.
Ritchie, J.
1912 An Account of the Watch-Houses, Mortsafes, and Public Vaults in Aberdeenshire Churchyards, Formerly Used for the Protection of the Dead from the Resurrectionists. *Proceedings of the Society of Antiquaries of Scotland* March 11, 1912:285–326.
Savitt, T. L.
1982 The Use of Blacks for Medical Experimentation and Demonstration in the Old South. *The Journal of Southern History* 48(3):331–348.

Shultz, S. M.

1992 *Body Snatching: The Robbing of Graves for the Education of Physicians in Early Nineteenth Century America.* McFarland, Jefferson, North Carolina.

Singer, C.

1957 *A Short History of Anatomy and Physiology from the Greeks to Harvey.* Dover, New York.

Tubbs, R. S., M. M. Shoja, M. Loukas, and P. Agutter

2019 *History of Anatomy: An International Perspective.* Wiley Blackwell, Hoboken, New Jersey.

Tward, A. D., and H. A. Patterson

2002 From Grave Robbing to Gifting: Cadaver Supply in the United States. *The Journal of the American Medical Association* 287(9):1183.

Wilson, J.

1998 *Stanford University School of Medicine and the Predecessor Schools: A Historical Perspective.* Lane Medical Library, Stanford University, Palo Alto, California.

3

The Role of US Army Surgeons and Their Contributions to the Army Medical Museum in the Post–Civil War Era (1865–1890)

BRIAN F. SPATOLA AND KRISTEN E. PEARLSTEIN

The Army Medical Museum (AMM) was founded in 1862 as part of a wave of reforms designed to address the inadequate state of military medicine at the start of the US Civil War. In June of 1861, federal legislation created the United States Sanitary Commission (USSC) to make recommendations to the government on how to improve health conditions in the Union Army. In April of 1862, the USSC recommended the appointment of William A. Hammond as the Army Surgeon General (Bayne-Jones 2007 [1968]; Duncan 1987) and recommended reorganization of the entire US Army Medical Department in order to meet the challenges of evacuating casualties, treating the injured, and preventing avoidable diseases such as typhoid and dysentery. The creation of the AMM was part of a larger plan to revolutionize Army medicine through detailed reporting and evaluation of the efficacy of medical care in order to improve the health outcomes of US Army soldiers on future battlefields. The establishment of the AMM was a pivotal moment in US history that initiated the mission to study the effectiveness of treatments for the injuries and diseases relevant to military forces.

The AMM was established by the order of Surgeon General William Hammond. On May 21, 1862, Hammond issued Circular No. 2, which established the museum and laid out the collecting agenda that would carry it far beyond the Civil War. In addition to completing the details of their monthly reports of sick and wounded, medical officers were also directed "to collect, and to forward to the office of the Surgeon General, all specimens of morbid anatomy, surgical or medical, which may be regarded as valuable; together with projectiles and foreign bodies removed and other matters as may prove of interest in the study of military medicine or surgery" (Hammond 1862). This material and the accompanying case histories supplied by the contributors were used in compiling the multivolume *Medical and Surgical History of the War of the Rebellion,* the last volume of which was published in 1888 (Otis and Huntington 1991 [1883]). This six-part publication contained summaries and statistical tabulations of medical reports received from the field, detailed case studies with treatment notes and patient outcomes, and documentation of the activities of the Army Medical Department.

The museum's success would not have been possible without the participation of Army medical officers. Donations by individual medical officers in the field were critical to the museum's efforts to improve the health outcomes for the treatment of soldiers. In the early years of the Civil War, medical officers were ordered to submit specimens, projectiles, and detailed medical histories to the museum on top of their grueling roles performing amputations and treating injuries and diseases. By the end of the war, the museum was a vast repository of surgical specimens consisting largely of injured and amputated bones, and medical specimens of pathological conditions related to camp life including typhoid and dysentery. Surgeons also donated weapons, projectiles, and examples of medical technology. With the accompanying surgeon reports, the museum became a central clearing house for reporting injuries and diseases of soldiers and the outcomes of their treatments.

By 1866, surgeons had contributed 4,719 surgical specimens almost exclusively from the Civil War (Woodhull 1866). After the war, contributions to the AMM steadily increased from medical activities at remote forts in the western US and from expeditions and battles with Native Americans. The collecting agenda was broadened after the war to include several other areas of medical and scientific interest, such as surgical instruments, and normal human and animal specimens. These changes were in line not only with military medical interests but also the complimentary anatomical and pathological interests exemplified by medical museums of Europe (Billings 1888).

Post–Civil War Military Medicine and Museum Activities

After the conclusion of the Civil War, military interests refocused on the American West. The context of large-scale treatment of battlefield injuries and camp diseases gave way to optimizing patient care at military barracks and hospitals in all areas of the country. The AMM continued to work to advance military medicine, and surgeons were still expected to make extensive reports to the Surgeon General and donate specimens with detailed histories (Billings 1870, Billings 1875, Otis 1871).

Changing environs and circumstances meant Army surgeons were deployed by the Army Medical Department to approximately 200 military posts constructed across the US, many in remote western territories. It was there that Army medical officers provided medical services to troops embroiled in the various battles and skirmishes of the Indian Wars. These assignments often took place more in the most austere of conditions. The environments where they worked were often remote from the established cities and towns of the eastern US and were characterized by infrequent supply deliveries, harsh weather conditions, the physical demands of grueling expeditions, Native American attacks, and an overall lack of proper sanitation and facilities. Woodward wrote in 1876, "... the troops stationed on the Indian frontier are most of the time exposed to all the hardships and privations of a state of actual war" (Woodward 1876a:4).

Depending on the size of the post, physician positions could be filled by a single post surgeon or include one or more assistant surgeons (Gerstle Smith 2013; Gillett 1989). Point San Jose is an example of a relatively small post located on the northern tip of the San Francisco Peninsula between the city and a larger installation, the Presidio (see Bright, this volume). From Point San Jose, then Assistant Surgeon Edwin Bentley (see Gavette, this volume) reported on local conditions that impacted the health of soldiers (Billings 1870; Billings 1875). Bentley also collected specimens from soldiers and civilian patients he treated from the local community, thereby contributing to the efforts of the AMM to continue the ongoing work of improving military medicine. Assistant Surgeons William Bradley and E. J. Marsh were also assigned to Point San Jose for a time, and likewise contributed to these efforts (Billings 1870; Billings 1875; Otis 1871).

The Medical Department also utilized the services of contract physicians, classified as acting assistant surgeons, to help staff frontier garrisons. These surgeons were primarily civilian doctors hired for fixed terms to treat military and military dependent patients (Tate 1999). Most senior medical officers, such as Bentley, were veterans of the Civil War and had experience treating traumatic and ballistic injuries with surgery and amputation. Veteran surgeons were also more familiar than their civilian counterparts with the bone and tissue infections that often accompanied penetrating injuries.

Surgeons and assistant surgeons regularly rotated between Army garrisons every few years to ensure that surgeons did not work too long in isolation and that posts did not continuously suffer from inexperienced medical staff (Gerstle Smith 2013). For example, Assistant Surgeon Carlos Carvallo served at Fort Union, New Mexico, from October 1876 until May 1878, and by 1880 was reassigned to Fort Laramie, Wyoming. Assistant Surgeon Alfred Conrad Girard served at Fort Keogh, Montana Territory, from 1878 to 1884. Assignments and promotions among Army physicians were tied to successful completion of written exams, established by Surgeon General Joseph K. Barnes to prove and maintain adequate medical knowledge and skill advancement. Assistant surgeons entered service at a rank equal to that of a first lieutenant and were eligible for promotion and testing after three years (Gillet 1989; Tate 1999). However, surgeons rarely advanced above the rank of major due to Congressional restrictions on Army size and reductions in honorary promotion for meritorious service (Gillett 1989; Woodward 1876a).

In the post–Civil War period, an Army surgeon's scientific work in the fields of medicine, anatomy, and anthropology was facilitated by the privileged social position they held as educated physicians and expert authorities. Their assignments took them into geographical, ecological, and cultural regions of the country then largely untouched by scientific exploration. The AMM and the Smithsonian Institution both encouraged medical officers to make specimen donations to expand the holdings of federal museums and establish the foundations of the earliest scientific collections of the country (Tate 1999). Ashburn writes, "Many medical officers were naturalists of one sort or another, botanists, ornithologists, mammalogists, ichthyologists, etc. Their contributions to knowledge of flora and fauna of the United States were numerous and important" (Ashburn 1929:127). Army surgeons made substantial contributions to the pursuit of scientific knowledge, particularly in the field of biology (Richmond 1918; Shufeldt 1923) and in their donations of biological and medical specimens to the AMM, the Smithsonian's US National Museum, and similar institutions. During this era, the AMM and the Smithsonian Institution were in neighboring buildings on the National Mall in Washington, DC, a fact that facilitated shared interests and substantial trade in items and specimens between the two institutions. In an instance of emphatic support for a medical officer's scientific contributions, the Surgeon General assigned Assistant Surgeon Elliot Coues to classify specimens for the Smithsonian that he had collected during his military service in Mississippi (Gillett 1995).

The Surgeon General's Office issued circulars and letters to medical officers detailing the expansion of the museum's interests and the role they were expected to play. Surgeon General Barnes and AMM Curator George Otis astutely recognized that the medical officers' new environs offered opportunities to expand the museum into new medical, anatomical, and anthropological frontiers (Barnes 1870). In addition to prioritizing surgical reports and donations of specimens originating from treatment of soldiers, the collecting agenda expanded in April 1867 to include animal pathology, poisonous insects and reptiles, Native American skeletons, and Native technology related to medicine and warfare (Lamb 1917). The AMM was the principal federal institution for the study of human skeletal variation before the establishment of the Department of Physical Anthropology at the US National Museum in 1904. In his report on the AMM's participation in the International Exposition of 1876, Otis summarizes the initiatives for collecting Native American materials, writing, " . . . Army medical officers in remote posts who had infrequent opportunities of contributing to the pathological material of the museum, but earnestly shared in the general desire of members of the corps to promote its welfare, forwarded donation of Indian crania, of specimens of natural history, and of objects of ethnological or archaeological interest" (Otis 1876:3).[1] In September 1888, Surgeon General John Moore expanded the scope of expected museum donations to include embryos, pregnant uteri, complete human skeletons, animal skeletons, and nearly every major organ system (Lamb 1917). The resulting specimens were summarized in occasional catalogs and distributed by the museum to academic institutions (Otis 1888; Yarrow 1876).

Living and Working Conditions, and Duties of Surgeon

The role of the Army surgeon as a contributor to the AMM should be seen in the context of the surgeon's larger role, often as the sole medical practitioner, sanitation inspector, and health officer for an entire garrison of soldiers and neighboring civilians. Between the end of the Civil War (1865) and Wounded Knee (1890), the last major battle of the Indian Wars (Gillett 1995), Army surgeons' struggles to manage the outbreak of infectious diseases on Army posts amounted to little more than administering the smallpox vaccine, rigid adherence to standards of hygiene, and applied common sense. Likewise, the manner in which they treated major injuries in the years before antiseptic techniques was often more conservative than during the Civil War. Amputations were less frequently undertaken to avoid ensuing infection. Until the science of germ theory became commonly accepted in the 1870s–1880s, medical officers had very little reason to alter their manner of treatment.

The typical soldier under the surgeon's care was subjected to arduous physical labor in difficult conditions and was exposed to outbreaks of diseases, such as cholera, yellow fever, and typhoid (Gillett 1995). The grave consequences of epidemics with regard to the effectiveness of the fighting force meant that sanitation and hygiene duties increasingly became the responsibilities of the post surgeon. His administrative duties multiplied significantly during this time period. The factors of crowding, poor sanitation, and "bad ventilation," which were all believed to contribute to the spread of cholera, were problems for the post surgeon to solve.

The subjects of air and water quality and hospital ventilation were critical concerns of the medical command before the rise of germ theory and the establishment of bacteriology as a science. To mitigate these issues, the surgeon became increasingly involved in the design of hospitals and in garrison sewage and plumbing projects. A considerable amount of a post surgeon's time was spent on preventive medicine activities, such as administering smallpox vaccinations, testing garrison water supplies, and overseeing the soldier's dietary needs and hygiene practices. Despite their best efforts, unsuitable conditions often persisted due to inadequate supplies and labor (Gillett 1995).

Outbreaks of cholera and yellow fever caused fluctuations in disease rates that Army surgeons treated. While smallpox immunization was mandatory for the Army when supplies were available, the civilian anti-vaccination movement at the time often threatened the health of nearby soldiers. The preferred and most effective preventive measure for mitigating the spread of infectious diseases was quarantine. Surgeons were also responsible for monitoring the quality of water supplied to garrisons and for inspecting feces of soldiers in instances when a soldier had more than two bowel movements within a 24-hour period. Before the detection of bacteria was possible, surgeons sent water samples for testing to the AMM laboratory. However, the surgeon might only receive a report on select chemicals detected, which was effectively useless for medical decision-making purposes. With the advent of the laboratory-based science of bacteriology, testing of samples had improved to the point where cholera ceased to be a major issue (Gillett 1995).

The accommodations that a surgeon might find awaiting him at a new assigned location varied considerably from post to post. The decade after the Civil War was one of rebuilding, repairing, and improving structures used to provide medical services. As a typical example, Assistant Surgeon William N. Notson arrived at Fort Concho, Texas, in January of 1868 to find medical facilities consisting of nothing more than a series of tents and a foundation for a single structure. To add to the difficulties he encountered, medical supplies for Fort Concho were delivered only twice a year from San Antonio, some 230 miles away. In contrast, by 1870, Fort Richardson, Texas, had a modern two-story hospital built by civilian architects with two wards, a dispensary, a main office, steward's room, kitchen, and dining hall (Tate 1999). In his address to the International Medical Congress of 1876, Woodward had this to say about medical officers stationed out West: "Those of you who practice medicine in cities, and enjoy all the stimulus of converse with intelligent fellow-laborers and of competition for success, can hardly realize the difficulties under which most of the work of these medical officers is done, at remote and usually isolated stations, or amidst the hardships and perils of campaigns in the wilderness against the Indians, with no special stimulus to investigation except the love of science and desire for usefulness" (Woodward 1876a:11).

Medical and Surgical Cases from Soldiers and Civilians out West

Specimens related to the treatment of soldiers and civilians in western territories reflect a significantly different working environment for the Army surgeon than that of the Civil War. Civil War injuries consisted primarily of gunshot wounds sustained in battle, while combat-related injuries in the West were often from smaller conflicts with Native Americans. More injuries were the result of transportation accidents, accidental discharges of weapons, or suicides by firearm. Additionally, many reports and donations came from treatments performed on civilian settlers who had limited access to medical care.

Veteran surgeons from the Civil War were familiar with the 1862 Circular No. 2 issued by Surgeon General Hammond and were experienced with the process of submitting specimens and associated reports to the Surgeon General's Office for inclusion in the collections of the AMM. After the war, members of the Army Medical Department were expected to continue providing specimens to the AMM for the purpose of accumulating "material such as is often needed for purposes of comparison in arriving at the intelligent interpretation of an individual case" (Woodward 1876a:9). Surgeons were sometimes contacted directly to send a specimen to the AMM if their report piqued the interest of the museum staff. A March 20, 1880, letter from AMM Curator George Otis to the Post Surgeon of Angel Island and Medical Director of Presidio, San Francisco, California, reads, "Sir–I have the honor to request that if any specimen was preserved in the case of Private Thomas Jones, K, 4th Army, recorded on the December 1879 report of Angel Island, California, it be forwarded to the Army Medical Museum" (Otis 1880). Army surgeons were often eager to respond to these requests and contribute specimens to the museum collections. A May 14, 1880, note from Assistant Surgeon Carlos Carvallo, stationed at Fort Laramie, Wyoming, reads, "In compliance with request in letter of the 6th inst [May 6] will make zealous efforts to augment the A.M.M. collection" (Carvallo 1880).

Circulars issued by the Surgeon General's Office directed and summarized the results of these collecting activities. Circular No. 3, issued in 1871 and titled *A Report of Surgical Cases Treated in the Army of the United States from 1865 to 1871,* reported, "The returns of the Army for the past five years present a comparatively large number of gunshot wounds, received either in Indian hostilities, or accidently, or in brawls, or attempts to evade or escape from arrest. The more important of those cases will be enumerated, as nearly as may be in the order of the anatomical regions in which the wounds were inflicted, with such subdivisions as are most convenient" (Otis 1871:6). Otis summarized 1,037 surgical cases from 1865–1871 following the same format as that used for the *Medical and Surgical History of the War of the Rebellion.* Table 3.1 shows the summary of wounds, accidents, and injuries Otis reported by compiling the monthly reports of the sick and wounded forwarded to the Surgeon General between July 1, 1865, and December 31, 1870 (Otis 1871).

Otis found that surgical reports of major operations, such as amputations, were largely favorable including a 38.5 percent mortality rate for thigh amputations contrasted with 53.8 percent mortality reported during the Civil War for this same operation (Otis 1876; Otis and Huntington 1991[1883]). These findings contrast sharply with the reluctance of surgeons to perform amputations noted earlier. On the improvement of amputation results from 1865 to 1871, Otis noted that surgeons "greatly profited by the experience of the war" where they performed with little regard to textbook instruction and rather preferred to "adapt their incisions to the exigencies of particular cases" (Otis 1871:218).

When armed conflict occurred, it was in the campaigns of the Indian Wars, as the US Army sought to clear and control occupied Native Ameri-

Table 3.1. Report of accidents and injuries, 1865–1871

Classification	Total
Burns and Scalds	2,003
Contusion	23,651
Concussion	152
Drowning	198
Sprains	13,731
Dislocations	635
Simple Fractures	1,380
Compound Fractures	219
Gunshot Wounds	3,213
Incised Wounds	6,774
Lacerated Wounds	3,038
Punctured Wounds	1,591
Poisoning	683
Other Accidents and Injuries	3,683
Homicide	68
Suicide	84
Hanging	2

Source: Otis 1871.

can territories. Orders were given for medical officers to forward a list of causalities after every engagement, including details regarding the context and nature of all injuries. Army surgeons accompanied troops into the field to treat the sick and wounded. Assistant Surgeon Edwin Bentley accompanied troops and commanded the Camp Supply field hospital at Tule Lake, California, during parts of the Modoc War of 1872–1873. In April and May of 1873, he operated on several men who were injured in engagements with the Modoc Indians and required the amputation of a limb, then sent the specimens and associated medical reports to the AMM for curation.

In the absence of a sustained armed conflict after the Civil War, the specimens collected in the subsequent decades reflected the hardships and isolation of a frontier post assignment. As an example, reports from Fort Davis, Texas, indicated that traumatic injuries were not overly common and were rarely due to enemy hostilities. Between 1869 and 1872, Assistant Surgeon Daniel Weisel reported treating only 12 gunshot wounds, all due to accidents or assaults (Clary 1972). Between 1880 and 1889, Army surgeons at Fort Davis performed only a handful of surgeries, including finger amputations due to baseball injuries, a bullet injury to the lower leg, and foot surgery due to inadequate shoes (Gerstle Smith 2013).

Annual reports issued by the Surgeon General's Office and compiled from data contributed by post surgeons across the country show that the numbers reported from Fort Davis were typical of the 1870s and 1880s. Examples of accidental gunshot injuries submitted to the AMM by Assistant Surgeons Francis H. Atkins and Augustus A. Yeomans include the amputated lower arm of a soldier who disembarked from a wagon with his hand over the muzzle of his rifle near Fort Larned, Kansas, in 1876 (Figure 3.1D), and the left knee and associated shrapnel from an exploded shotgun barrel at Fort Richardson, Texas in 1878 (Figure 3.1E).

In addition to gunshot injuries, surgical reports described cases of frostbite, arrow injuries, suicide, and fractures (Ashburn 1929; Otis 1871). Frostbite was a frequent problem during cold weather campaigns against Native Americans and required amputation to prevent the spread of gangrene. A specimen contributed to the AMM in 1871 consists of a mummified amputated foot exhibiting gangrene due to frostbite from a member of the 6th Calvary stationed at Fort Richardson, Texas (Figure 3.1C).

Arrow wounds were infrequently encountered, and few surgeons developed the technique necessary to remove the arrow without dislodging the head from the shaft (Gillett 1995). Several arrow-wound specimens submitted to the AMM during the Indian Wars were contributed by Assistant Surgeon Notson from within 100 miles of Fort Concho, Texas. One skull represents a soldier killed by a Comanche arrow while the soldier was acting as an escort to a mail coach near Fort Concho in 1870 (Figure 3.1A). Another cranium represents an unidentified Mexican killed 75 miles from Fort Concho in 1868 by an arrow entering just above the eye (Figure 3.1B). Arrow wounds had all but disappeared in the medical literature with the use of firearms in warfare and these new cases provided a renewed interest in the subject. In particular it was noted that due to the rapidity that Native Americans discharged arrows, injuries were more often than not multiple rather than just a single arrow wound (Otis 1871). A skilled warrior could reportedly release up to six arrows per minute (Mays et al. 1994).

Eleven cases of death by suicide were reported to the Surgeon General between 1866 and 1871 with detailed descriptions of the cause, manner, and injury. Assistant Surgeon William Bradley reported from Point San Jose, California, on the September 23, 1868, death of Private Octavius E. Daniels by self-inflicted musket wound. A description of the injury was made without an autopsy (Otis 1871). A specimen submitted to AMM in 1896 from Fort Douglas, Utah, by Alfred Conrad Girard consists of a calvarium showing multiple gunshot fractures from a self-inflicted rifle (Krag-Jorgensen)

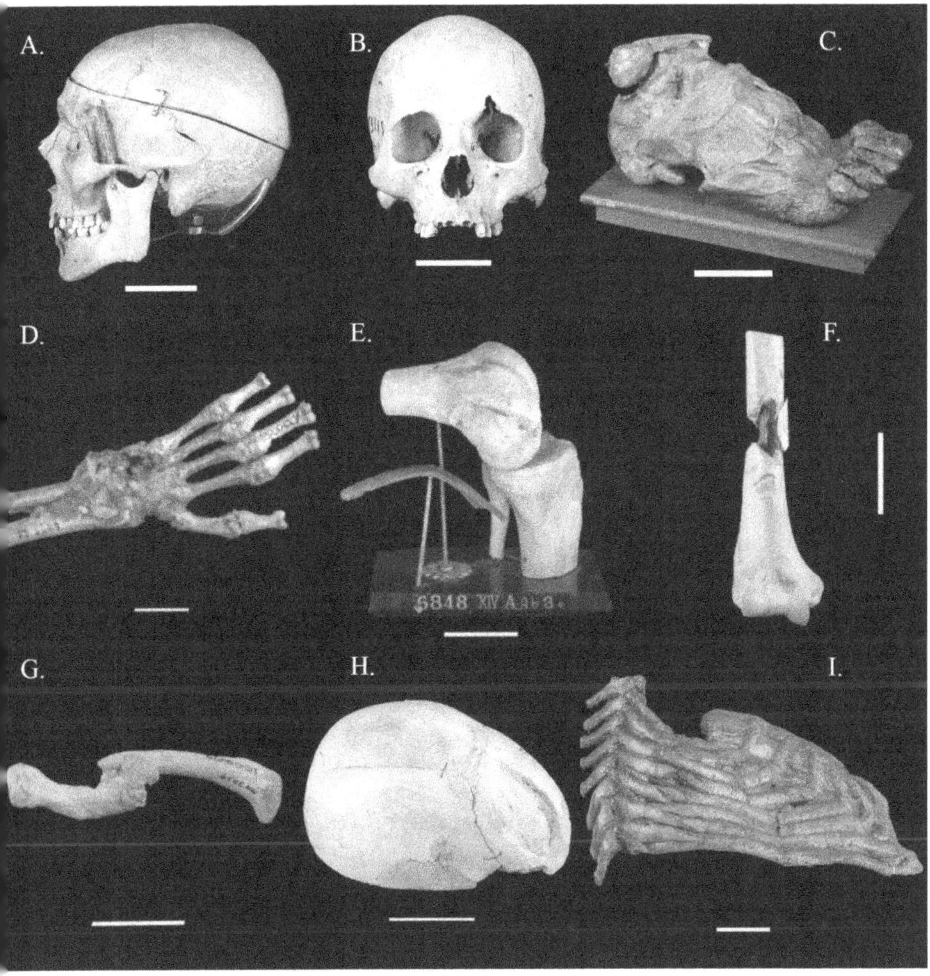

igure 3.1. Examples of specimens contributed to the Army Medical Museum between 1870 and 889 (scale equals 5 cm). (*A*) Skull with arrow injury to left temporal (PS 5908); (*B*) Cranium with rrow injury above left orbit (AFIP 551183); (*C*) Amputated right foot with frostbite and gangrene AFIP 1001905); (*D*) Amputated lower right arm with accidental gunshot injury (AFIP 1000052); *E*) Left knee with accidental rifle injury (AFIP 1001838); (*F*) Amputated right distal humerus with ircular saw injury (AFIP 0384992); (*G*) Clavicle with healed fracture (AFIP 1002175); (*H*) Cal-arium with healed depressed fracture (AFIP 1002490); (*I*) Vertebrae and fractured right ribs (AFIP 001812). Courtesy of National Museum of Health and Medicine, Silver Spring, Maryland.

wound. Girard's accompanying report reads, "There is a history of an injury to the head, while he was in Troop A, 6th Cavalry, at Fort Niobrara in 1892, received from a kick of the horse, since which time his disposition was altered—while formerly he was lively and enjoyed himself with the men, he became morose, depressed, secluded himself, expressed suicidal impulses" (Girard 1896). These specimens and reports were important contributions to the AMM's ongoing research into the treatment of maladies of soldiers through documentation of projectile injury patterns from newer weapons and munitions, tabulation of surgical outcomes, and for documenting the signs and symptoms of a traumatic brain injury.

Fractures resulting from accidents were common, and a review of the 1871 Circular No. 3 indicates that many fractures were received due to accidents, falls and brawls following alcohol consumption, proximity to horses, and proximity to items of transportation, including wagons and trains. An example of a sharp-force injury accident consists of an amputated distal right humerus with cut marks from contact with a circular saw (Figure 3.1F). Musician Frank Horning, Troop A, 7th Infantry, was caught on the arm by a steam-driven circular lumber saw at Fort Ellis, Montana Territory, in February 1872. After the loss of his arm, the 19-year-old was transferred to the Soldiers' Home, Washington, DC, for retirement (Campbell 1872). In Arizona Territory, Acting Assistant Surgeon George H. R. Moran wrote in his diary in February 1878 that various sick calls included one man "run over by a wagon and ankle sprained" and one man "whose head was cut by stage [coach] upsetting as it left station" (Moran and Hagemann 1963).

Specimens contributed to the AMM often represented healed fractures from prior injuries that were discovered during autopsy following an untimely death. Examples of healed fractures from horse-related injuries include a clavicle with a healed fracture resulting from being thrown from a horse near Fort Griffin, Texas, contributed by Carlos Carvallo in 1872 (Figure 3.1G). The soldier recovered and returned to duty six weeks after the injury but died one month later of cerebral meningitis (Carvallo 1872). A second example consists of the calvarium of Private Francis Young, Troop B, 10th Cavalry, exhibiting a large healed depressed fracture of the frontal (Figure 3.1H). Private Young was on duty as a teamster near Fort Marcy, Virginia, in November 1879 when his team fell down the side of a mountain, drawing him with them. He was found nearly strangled by a wagon chain and with contusions of the head and neck (Collins 1879). Private Young recovered from those injuries and returned to duty, but he died 10 years later in May 1889 following a second traumatic head injury when he was thrown from a horse and landed directly on his head near Fort Apache, Arizona Territory (Everts 1889). A different transportation accident consists of the partial vertebral column and fractured right ribs of a civilian laborer who was run over by the wheel of a wagon carrying approximately 3,000 lbs. at Fort Sanders, Wyoming Territory, in 1870 (Figure 3.1I). The associated medical report describes obliteration of the pleural cavity and the direction of the turning wheel as it passed diagonally over the torso (Frantz 1870).

As major armed conflicts became less frequent, surgeons found themselves treating more civilians living and working around frontier Army posts. Army surgeons in proximity to settlers were often the only available physician. In this circumstance, the Surgeon General encouraged surgeons to establish private practices during off-duty hours so knowledge they gained treating local medical conditions could be applied toward improved care for soldiers (Gillett 1995). The rise of civilian medical care at Army outposts was reflected in the number of specimens submitted to the AMM from civilian patients. Woodward noted in 1871, "Since the war . . . the number of preparations which exhibit the morbid anatomy of the diseases of civil life is constantly increasing. . . ." (Woodward 1871:236). Surgeon General Barnes and AMM Assistant Surgeons George Otis and John Shaw Billings encouraged contributions from civilian patients as a means to broaden the scope of the museum collections with specimens of general medical interest (Gillett 1995; Lamb 1923). By 1876, while the primary collecting emphasis was still for specimens of wounds and diseases that caused death and disability in military forces, Woodward observed that the Surgeon General wished for the collection to be " . . . extended so as to embrace all forms of injuries and diseases, so that eventually it shall become a general pathological museum, accessible for study to all medical men who are prosecuting original inquires" (Woodward 1876b:3).

Army surgeons were not required by the federal government to treat their civilian neighbors, but responded to the growing demand nonetheless. Assistant Surgeon Notson explained in 1871, "' . . . humanity and custom, compels the Medical officer, not only to exert his professional ability, but to open his hospital for their care'" (Matthews 2005:54). As a result of the inclusion of non-military patients, surgeons had access to types of cases that did not readily fit into the usual categories of gunshots, fractures, and accidents. In 1870, Assistant Surgeon Edwin Bentley contributed a cranium and mandible with "syphilitic hyperostosis" from a 29-year-old San Francisco female of "depraved, intemperate habits, and kindred indulgences"

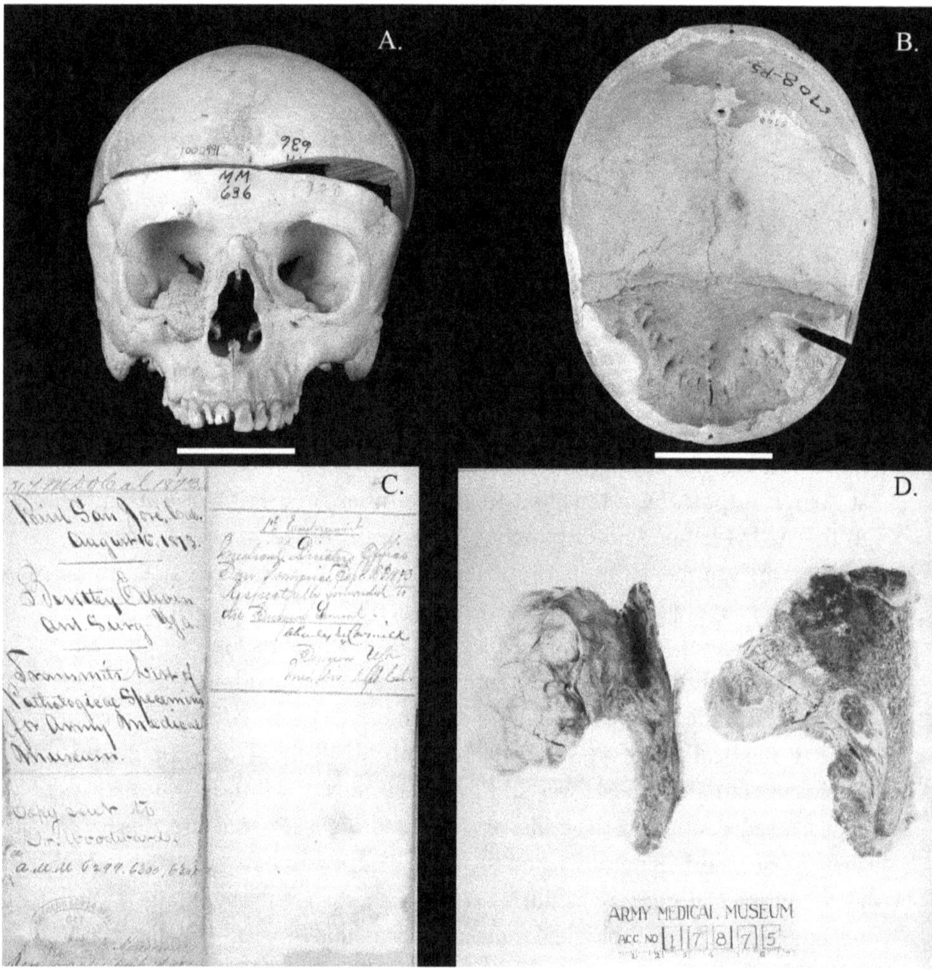

Figure 3.2. Examples of specimens and document contributed to the Army Medical Museum from Point San Jose, California, by Assistant Surgeon Edwin Bentley (scale equals 5 cm). (*A*) Cranium with hyperostotic bone masses (AFIP 1002491); (*B*) View of endocranial surface with hyperostotic bone mass (AFIP 1002491); (*C*) Document accompanying pathological specimens from Point San Jose, California, in 1873; (*D*) Image of partial scapula with enchondroma (AFIP 0017875). Courtesy of National Museum of Health and Medicine, Silver Spring, Maryland.

(Bentley 1870). The cranium shows no evidence of a treponemal infection; rather, the hyperostotic areas are two distinct masses of dense new bone, one originating from the right maxillary sinus extending into the eye orbit (Figure 3.2A), and the other formed along the endocranial surface of the frontal bone (Figure 3.2B). The latter is similar in appearance to hyperostosis frontalis interna with some key differences, notably the young age of the patient, the cortical-like density of the frontal mass in cross section, and the consistency of density between the two masses.

In August 1873, Assistant Surgeon Bentley contributed the scapula of a civilian farrier from Point San Jose, California, that exhibited an enchondroma, a non-malignant bone tumor (Figure 3.2D). Assistant Surgeon Bentley augmented the AMM collections with two more civilian specimens from Point San Jose in August 1873 (Figure 3.2C): a fractured cranium and a necrotic knee joint. Contributions of unusual cases, such as these specimens from Assistant Surgeon Bentley, assisted the growing museum in establishing a reference collection that met the needs of the medical community beyond the injuries common in a military population. Contributions of civilian specimens greatly decreased after 1892 following an order from Surgeon General Sutherland for Army surgeons to discontinue treating private citizens and an 1893 congressional budgetary directive that eliminated or reduced the number of contract and assistant surgeon positions in the Medical Department.

Summary and Discussion

The working conditions, responsibilities, and nature of specimen donations from medical officers during the Indian Wars reflect several important differences from those of the earlier Civil War. Skirmishes and battles with Native Americans were sporadic and the nature of the injuries reflects the technological adaptations and practices of Native Americans, namely gunshot injuries or multiple wounds from arrows with iron points. The incidence of injuries from accidents greatly increased compared to those from gunshot wounds received during warfare. In addition to their duties treating soldiers, surgeons acted in diverse administrative and practical roles including sanitation, hygiene, and facilities planning. The surgeon as a physician became increasingly involved with treating neighboring civilian populations, commencing a long history of synergy between military and civilian medicine.

The specimens from the Indian Wars era collected by the AMM demonstrate several differences from those of the Civil War. Arrow injuries facilitated a renewed interest in the treatment of such wounds that had not been seen regularly since the advent of modern firearms. Numerous transportation accidents from horse-drawn vehicles and from railroads began to receive regular treatment from Army doctors. Suicides and accidents involving new models of firearms, such as the Krag-Jorgensen rifle, occurred. Surgical and autopsy examples were forwarded with the intent that such an accumulation of cases would lead to recommendations on the treatment of similar injuries in the future. The submission of cases from the Indian Wars is part of a long history of similar studies of battlefield medicine that continues to this day.

The Museum Today

Today, the AMM is renamed the National Museum of Health and Medicine. The museum evolved to become the primary repository and archive for the research and educational activities of military medicine in the contexts of medical history, anatomy, and pathology. Since its inception, the museum has passed through several iterations, influenced by the changing technological, scientific, and research interests of US military and civilian medicine. Today, the museum is a unique stand-alone military medical asset of the Department of Defense and the premier medical museum of the US.

Disclaimer

The opinions and assertions contained herein are the expressed views of the authors and are not to be construed as official or reflecting the views of the Department of the Army, Navy, or Air Force; of the Department of Defense; or the Defense Health Agency.

Note

1. The role of the AMM in collecting Native American skeletons, the desecration of Native American graves, and opportunistic collecting of Native American artifacts has been discussed and critiqued in numerous publications (Redman 2016; Thomas 2000).

References Cited

Ashburn, Percy Moreau
1929 *A History of the Medical Department of the United States Army.* Houghton Mifflin, Boston.

Barnes, Joseph K.
1870 *Annual Report of the Surgeon General, United States Army.* Surgeon General's Office, Washington, DC.

Bayne-Jones, Stanhope
2007 [1968] The Evolution of Preventative Medicine in the US Army, 1607–1939. Office of the Surgeon General, Dept. of the Army, Washington, DC. Borden Institute, US Army. Reprint.

Bentley, Edwin
1870 A Remarkable Specimen of Puerperal Osteophytes. *Pacific Medical and Surgical Journal* 4:1–5.

Billings, John S.
1870 *Report on Barracks and Hospitals with Descriptions of Military Posts.* Government Printing Office. Washington, DC, Circular #4, Dec. 5, 1870, War Department Surgeon General's Office.
1875 *A Report on the Hygiene of the US Army with Descriptions of Military Posts.* Government Printing Office. Washington, DC, Circular #8, May 1, 1875, War Department Surgeon General's Office.
1888 *Medical Museums, with Special Reference to the Army Medical Museum at Washington: The President's Address, Delivered Before the Congress of American Physicians and Surgeons, September 20, 1888.* Tuttle, Morehouse and Taylor, New Haven, Connecticut.

Campbell, Archibald B.
1872 Catalog Card. Accession File, AFIP 0384992, OHA 33, Otis Historical Archives, National Museum of Health and Medicine, Silver Spring, Maryland.

Carvallo, Carlos
1872 June 17. Untitled medical report. Accession File, AFIP 1002175, OHA 33, Otis Historical Archives, National Museum of Health and Medicine, Silver Spring, Maryland.
1880 May 14. Letter summary regarding donation. Accession File, AFIP 1002014, OHA 33, Otis Historical Archives, National Museum of Health and Medicine, Silver Spring, Maryland.

Clary, David A.
1972 The Role of the Army Surgeon in the West: Daniel Weisel at Fort Davis, Texas, 1868–1872. *Western Historical Quarterly* 3(1):53–66.

Collins, J. H.
1879 Untitled medical report. Accession File, AFIP 1002490, OHA 33, Otis Historical Archives, National Museum of Health and Medicine, Silver Spring, Maryland.

Duncan, Louis C.
1987 [191-?] The Medical Department of the US Army in the Civil War. Reprint by Olde Soldier's Books, Gaithersburg, Maryland.

Everts, Edward
1889 May 31. *Record of Case and Post Mortem Examination of Private Francis Young, Troop "B," 10th Cavalry.* Accession File, AFIP 1002490, OHA 33, Otis Historical Archives, National Museum of Health and Medicine, Silver Spring, Maryland.
Frantz, J. H.
1870 May 2. *Notes relative to specimen for Army Medical Museum.* Accession File, AFIP 1001812, OHA 33, Otis Historical Archives, National Museum of Health and Medicine, Silver Spring, Maryland.
Gerstle Smith, Donna
2013 Till Death (or Discharge) Do Us Part: The State of Medicine at the Nineteenth Century US Army Post of Fort Davis, Texas. *Journal of Big Bend Studies* 25:105–117.
Gillett, Mary C.
1989 United States Army Surgeons and the Big Horn-Yellowstone Expedition of 1876. *Montana: The Magazine of Western History* 39(1):16–27.
1995 *The Army Medical Department 1865–1917.* Army Historical Series, Center of Military History, United States Army, Washington, DC.
Girard, Alfred Conrad
1896 March 11. *Suicidal Gunshot Wound of the Head with the Modern Service Rifle.* Accession File, AFIP 1001045, OHA 33, Otis Historical Archives, National Museum of Health and Medicine, Silver Spring, Maryland.
Hammond, William A.
1862 Circular No. 2. Washington City, Surgeon General's Office, Washington, DC.
Lamb, Daniel S.
1917 A History of the United States Army Medical Museum 1862 to 1917: Compiled from the Official Records. Manuscript on file, Otis Historical Archives, National Museum of Health and Medicine, Silver Spring, Maryland.
1923 The Army Medical Museum, Washington, DC. *The Military Surgeon* 53(2):89–140.
Matthews, James T.
2005 *Fort Concho: A History and a Guide.* Texas State Historical Association, Austin.
Mays, B., A. Parfitt, and M. J. Hershman
1994 Treatment of Arrow Wounds by Nineteenth Century USA Army Surgeons. *Journal of the Royal Society of Medicine* 87:102–103.
Moran, George H. R., and E. R. Hagemann
1963 Arizona Territory–1878: The Diary of George H. R. Moran: Contract Surgeon, United States Army. *Arizona and the West* 5(3):249–267.
Otis, George A.
1871 *A Report of Surgical Cases Treated in the Army of the United States from 1865 to 1871.* Circular No. 3, Surgeon General's Office, War Department. Government Printing Office, Washington, DC.
1876 *Check List of Preparations and Objects in the Section of Human Anatomy of the US Army Medical Museum for Use during the International Exhibition of 1876 in Connection with the Medical Department, US Army.* No. 8, Army Medical Museum, Washington, DC.

1880 March 20. Letter to Medical Director, Department of California, Presidio, San Francisco, California. Curatorial Letterbook 1869–1886, OHA 15, Otis Historical Archives, National Museum of Health and Medicine, Silver Spring, Maryland.

1888 *List of the Specimens in the Anatomical Section of the US Army Medical Museum.* Gibson Brothers, Washington, DC.

Otis, George A., and D. L. Huntington

1991 [1883] The Medical and Surgical History of the War of the Rebellion, Part III, Vol. 2 Surgical History. 1991, second edition. Broadfoot Publishing, Wilmington, North Carolina.

Redman, Samuel J.

2016 *Bone Rooms: From Scientific Racism to Human Prehistory in Museums.* Harvard University Press, Cambridge, Massachusetts.

Richmond, Charles W.

1918 In Memoriam: Edgar Alexander Mearns. *The Auk* 35:1.

Shufeldt, Robert W.

1923 Elliot Coues: Naturalist and Surgeon. *Medical Life* 30:1.

Tate, Michael L.

1999 *The Frontier Army in the Settlement of the West.* University of Oklahoma Press, Norman.

Thomas, David H.

2000 *Skull Wars: Kennewick Man, Archaeology, and the Battle for Native American Identity.* Basic Books, New York.

Woodhull, Alfred A.

1866 *Catalogue of the Surgical Section of the United States Army Medical Museum.* Government Printing Office, Washington, DC.

Woodward, Joseph J.

1871 The Army Medical Museum at Washington. *Lippincott's Magazine* 7:233–242.

1876a *The Medical Staff of the United States Army, and Its Scientific Work: An Address Delivered to the International Medical Congress at Philadelphia.* Gibson Brothers, Washington, DC.

1876b Advertisement. In *List of Skeletons and Crania in the Section of Comparative Anatomy of the United States Army Medical Museum: For Use during the International Exhibition of 1876 in Connection with the Representation of the Medical Department, US Army,* by Henry C. Yarrow, pages 3–4. Collins, Philadelphia.

Yarrow, Henry C.

1876 *List of Skeletons and Crania in the Section of Comparative Anatomy of the United States Army Medical Museum: For Use during the International Exhibition of 1876 in Connection with the Representation of the Medical Department, US Army.* Collins, Philadelphia.

4

A Surgeon among Surgeons

The Medical Staff at Point San Jose

PETER GAVETTE

The discovery of a pit filled with human remains mixed with medical waste raises a number of questions. The first question asks why anatomized remains would be deposited in this fashion behind a military hospital and who could be responsible. Another important question asks where the remains came from and who they represented.

Setting aside the question of why the remains were disposed of in this manner, the location, adjacent to the Point San Jose post hospital, suggests that a surgeon or steward associated with the hospital was most likely responsible. Osteological conclusions (see Cox et al., and Sgheiza and Willey, this volume) have shown that the demographic profile is inconsistent with soldiers. Further, only 11 soldiers died at Point San Jose (Markland 2001) between 1864 and 1885, not enough to represent the minimum number of individuals (n=25) recovered from the pit (see Cox et al., this volume) during the estimated time that the associated artifacts indicate (1860s to 1870s, see Locke Barton, this volume).

From 1864 to 1902, when Building 239 served as the post hospital before it was relocated to the larger Building 201, 46 surgeons were assigned to Point San Jose (National Archives and Records Administration [NARA] 1825–1909:4–7). Given the nature of military service, surgeons were stationed for temporary terms, some much shorter than others. The average assignment at Point San Jose was 1.9 years (Figure 4.1). The shortest duration was seven days, by Surgeon Albert Hartsuff in June 1894. The longest term was more than five years by Assistant Surgeon Edwin Bentley, who was stationed at Point San Jose from August 31, 1869, through January 1, 1875. To begin to answer the question of who might be responsible, one

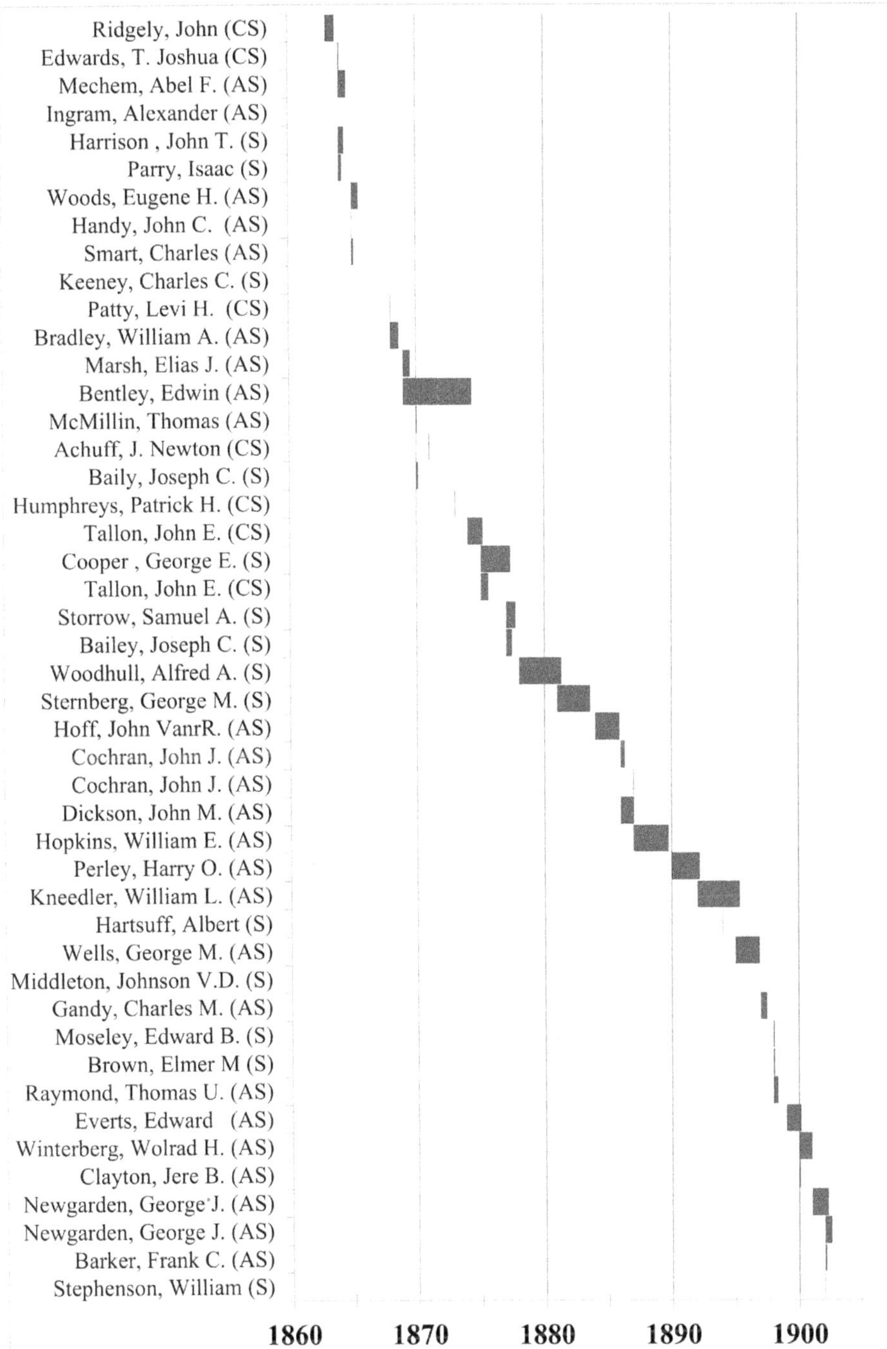

Figure 4.1. Surgeons (S), Assistant Surgeons (AS), and Contract Surgeons (CS) posted at Point San Jose from 1863 through 1902.

could look at the contributions of those who served at Point San Jose to the Army Medical Museum as analogue for their actions as surgeons.

The Army Medical Museum

Surgeon General William Hammond, in his May 21, 1862, Circular No. 2 (Hammond 1862), directed medical officers in the field to "collect, and to forward to the office of the Surgeon General all specimens of morbid anatomy, surgical or medical, which may be regarded as valuable." These collections were curated in the then newly founded Army Medical Museum (AMM; currently the National Museum of Health and Medicine) in hopes that they might yield insights to improve care of Union soldiers. Museum staff also visited battlefields and hospitals to collect specimens themselves (Brinton 1914, Rutkow 2005). The collections were divided into three categories: microscopic, surgical, and medical (for more information on the AMM, see Spatola and Pearlstein, this volume).

Of the surgeons who served at Point San Jose, six are listed as having contributed to the AMM during the Civil War. Assistant Surgeon Alexander Ingram contributed 15 surgical specimens. Ingram was not likely a long-term surgeon at Point San Jose; he started in August of 1864 with no end date listed. While some surgeons' service was concurrent, it was generally not for long, and in December of 1864, two new surgeons began their terms at Point San Jose, so it is likely that Ingram had left by then or shortly thereafter. Assistant Surgeon William Bradley, who was posted at Point San Jose for seven months, contributed four medical specimens and one surgical specimen. He was also there for the one surgical case reported from Point San Jose between 1865 and 1871 (Otis 1871), in which Private Octavius E. Daniels committed suicide by placing the muzzle of his musket against his temple on September 23, 1868. No autopsy was made. Surgeon George Sternberg, who contributed two specimens, was stationed at Point San Jose for two and a half years. Two other surgeons contributed six surgical specimens each: Assistant Surgeon Elias J. Marsh posted six months and Surgeon Samuel A. Storrow posted eight months at Point San Jose. Assistant Surgeon Edwin Bentley, in addition to being the longest serving surgeon at Point San Jose, was one of the most prolific contributors to the AMM during the Civil War, submitting 64 medical and 89 surgical specimens by 1866 (Woodhull 1866).

Undoubtedly, a surgeon would have needed a local and reliable source of cadavers to accumulate at least 25 individuals disposed in the Point San

Jose medical waste pit. Given these parameters, Bentley's tenure stands out, having served two years longer than the next closest surgeon. In addition, his prolific contributions to the AMM far outpaced his fellow surgeons who were posted at Point San Jose. These factors, along with others presented below, combine to reveal Bentley as the prime suspect as the source of the discarded human remains.

Bentley Biography

Edwin Bentley was born on July 3, 1824, in New London, Connecticut, to George W. and Annie (Williams) Bentley (Watson 1896, Cobb 1980; Figure 4.2). After his education in common schools and by private tutors, he commenced studying medicine in 1846 with Dr. Thomas P. Wattles of North

Figure 4.2. *Top left to bottom right:* Dr. Edwin Bentley during the Civil War, from the collection of Melinda and Paul R. Johnson; Bentley in uniform in San Francisco, ca. 1869; Bentley in civilian clothes in San Francisco, ca. 1875, from the U.S. National Library of Medicine; Bentley as professor at University of Arkansas Medical Department, 1897, courtesy of Historical Research Center, University of Arkansas for Medical Sciences; Bentley seated second row center during Modoc Campaign, 1873, courtesy of the National Archives and Records Administration; Bentley teaching at University of Arkansas Medical Department, 1903, courtesy of the Historical Research Center, University of Arkansas for Medical Sciences.

Stonington, Connecticut, and later under the preceptorship of Dr. John H. Whittaker, of New York City (Watson 1896). He attended three lecture courses at the University Medical College of New York City (now New York University School of Medicine), graduating in 1849 (Watson 1896). Bentley began medical practice in Norwich, Connecticut, soon after graduation. He also graduated with medical diplomas from the College of Physicians and Surgeons in New York City (now Columbia University Vagelos College of Physicians and Surgeons) in 1878 and *ad eundem* from Bellevue Hospital Medical College (1861–1898) in 1878 (Lamb: 1900; Watson 1896).

Civil War Period

After practicing private medicine in Connecticut for 12 years, the Civil War commenced, and Bentley mustered into service on June 6, 1861, in the Fourth Connecticut Infantry as an assistant surgeon (James T. White & Co. 1896; Powell and Shippen 1892; Watson 1896). He was then appointed surgeon of US Volunteers on September 4 of the same year (Patterson 1934). In January 1862, this unit became the First Connecticut Heavy Artillery, and in September of that year, Bentley was placed in charge of the Third Division US Army General Hospital in Alexandria, Virginia (Cobb 1980; Powell and Shippen 1892). Located across the Potomac River from Washington, DC, Alexandria was an important post and the center of medical care for Union forces. By 1864, he was in charge of all Alexandria-area hospitals, including L'Ouverture Hospital, dedicated to the treatment of African American and Native American soldiers (Lamb 1900; Watson 1896). In addition to soldiers, L'Ouverture treated African American civilians, many of whom had escaped slavery.

Medical and Surgical History of the War of the Rebellion

The contributions to the advancement of medical science during war is well-documented (Battle 1978; Blaisdell 1988; Guyatt 2001; Laffin 1999; Stansbury and Hess 2009). Even the ancient Greek physician and father of medicine Hippocrates wrote, "He who would become a surgeon should join an army and follow it" [translation by Potter (1995:315)]. Conflict and the resultant injuries, infectious diseases, and mass casualties offered plentiful patients on whom surgeons could hone their skills and test new methods.

Surgeon General Hammond issued an order on June 9, 1862, for Army surgeons to collect data on the sick and wounded from the battlefield and military hospitals. These medical data collected during the Civil War formed the foundation for the *Medical and Surgical History of the War of the Rebellion, 1861–65* (MSHWR) (Barnes 1870–1888). It consisted of six volumes issued between 1870 and 1888. The work was divided into three parts, each part consisting of medical history and surgical history volumes, totaling nearly 3,000 pages. The work details tens of thousands of surgical cases and diseases that occurred during the Civil War. The publication is acknowledged as the first major academic accomplishment in medicine in the US and became an early testament to evidence-based medicine (Bollett 2004).

Bentley responded to Hammond's order and was among only five brigade surgeons acknowledged for their ability and efficiency in the MSHWR. As indicated by the medical history of the rebellion, Bentley distinguished himself by "his great skill and the contributions which he made to medical and surgical literature of the rebellion, during which time he contributed more scientific articles than any other surgeon" (Altenberg et al. 1917).

Bentley's prodigious contributions are apparent in the index for each of the volumes, and he receives his own footnote in Part II of Volume I Medical History (Barnes 1879:193) regarding the nearly 100 autopsies submitted, stating, "It is to be regretted that, in most instances, the records of this hospital do not show by whom the autopsies were made. It is known that many of them were made by Surgeon Bentley himself, or under his immediate supervision, but it is only possible to distinguish these from the others in a few cases."

Among the many contributions Bentley made to the MSHWR, two stand out as testament to his skill as surgeon and innovation in medical practice. The case of Private George Lemon is considered one of the most remarkable cases documented in the MSHWR and was one of seven successful hip amputations illustrated in George Otis's 1867 tome on the subject. Hip amputation was among the most drastic surgeries known to medicine at the time. During the Civil War, only 66 such surgeries were performed, with a mortality rate of 83 percent (Cutter 2015). In 1867, two years after his leg was amputated, Lemon reported that his health was excellent and he was "able to walk to the village of Westminster, a distance of seven miles, without fatigue" (Barnes 1883:145). One specimen from the Point San Jose collection, Figure 4.3 (Catalog 42380) displays a saw cut through the neck of the femur. Another specimen, Figure 4.3 (Catalog 43043) represents a head removed from a femur, both possible attempts to practice or demonstrate this procedure.

There were two cases of human-to-human blood transfusion recorded in the MSHWR, one successful, and one fatal. The former, performed by

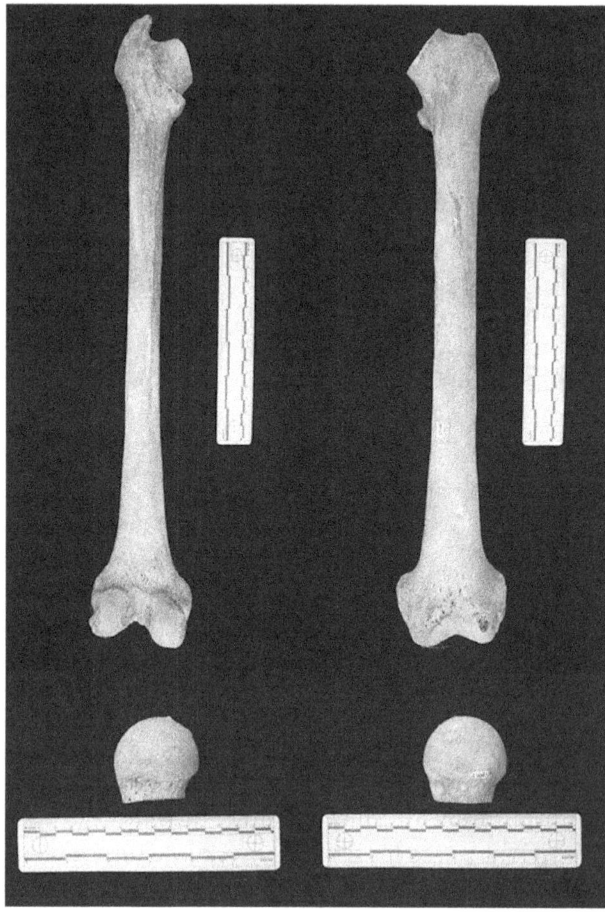

Figure 4.3. *Top to bottom*: Catalog #42380 exhibits a saw cut through the neck of the femur. Another specimen, Catalog #43043 represents a head removed from a femur. Courtesy of National Park Service and Chico State.

Bentley, is the first documented successful human-to-human blood transfusion in the US (Weld 1965, 1973). Bentley saved the life of Private G. P. Cross in August of 1864, after having been wounded in his right leg, by way of a transfusion of blood "from the temporal artery of a strong healthy German" (Barnes 1883:811).

Post–Civil War Period

On February 28, 1866, following the Civil War, Bentley was appointed assistant surgeon in the Regular Army, and was post surgeon at Russell Barracks, DC, until December 1868 (NARA 1888; Patterson 1934). He was then post surgeon at Lincoln Barracks, DC, until April 1869 (NARA 1888; Patterson 1934).

Howard University

While in Washington, Bentley became the first professor of anatomy at Howard University, a historically Black school. He held the position from 1868 until 1870 (Cobb 1980; Howard University 1869; Sampson 1967). With experience gained from serving during the Civil War, Howard recruited most of its medical faculty from the military ranks (Cobb 1980:1014). Bentley's instruction included "postmortem demonstrations" at Howard's Freedmen's Hospital (Lamb 1900).

Department of California

In April 1869 at 44 years old, Bentley was reassigned to the Department of California, one of two Army departments west of the Rocky Mountains and south of Oregon Territory (NARA 1888; Patterson 1934). He landed for a short stint at Camp Reynolds on Angel Island (NARA 1888; Patterson 1934; Map 2). That same month, the *Sacramento Daily Union* reprinted an article from the *Norwich Connecticut Bulletin* titled "A Lively Sensation in Norwich" (*Sacramento Daily Union,* 15 April 1869:1, emphasis added).

> There was an administrator's sale of personal effects at the former residence of the late Thomas Norris on Saturday. In the schedule of property was the following item: One fifteen gallon keg, contents unknown, 75 cents. No one seemed disposed to purchase the keg, and it was taken at the appraisal by John R. Dorrance & Co. Dorrance was unable to tell what fluid the keg contained. Thinking that all the value was in the keg itself, the fluid was allowed to run out on the ground for several hours in front of the store. When about closing for the night, the clerk tried to roll in the keg in question and was astonished to find that, from its weight, there was still something in it. To gratify his curiosity he procured a hammer and burst in the bead. To his astonishment a human hand and arm protruded. Without waiting for further developments, he fled in terror, and told every one [*sic*] he could find the cause of his fright. The news soon attracted a large crowd, and the contents of the package were still further examined. *Other arms, hands and legs, and two ghastly heads, were discovered.* Here was a mystery of startling proportions. Who could be the horrid monster who had committed such a crime and then so easily disposed of the evidences of it? And in Norwich? Ugh! Where were the authorities? Crowds rushed to the scene of the discovery. It soon came to the ears of some of the physicians in town, and they hurried

> to the scene of the excitement and solved the mystery. One of them recollected that the house of Norris was the property and the former residence of Dr. Edwin Bentley.

This article would portend Bentley's ensuing endeavors.

Remaining at Camp Reynolds from May through August, Bentley then moved to his post at Point San Jose, San Francisco, where he served as post surgeon (NARA 1888; Patterson 1934). While at Point San Jose, he examined recruits and attended the sick. He also cared for the sick on Alcatraz Island (NARA 1806–1902). Bentley engaged in the civic life of San Francisco, as well. In addition to his military duties, he was professor of descriptive and microscopic anatomy and pathology at the Medical College of the Pacific (Hempstead 1911; Watson 1896).

From April through June of 1873, Bentley accompanied the 4th Artillery to Tule Lake on the Modoc Campaign (NARA 1888; Patterson 1934; Watson 1896), where hostilities had erupted between Native Americans and settlers. He established a field hospital north of Tule Lake during the war and at its conclusion in April 1873 transported the wounded to Fort Klamath, Oregon (NARA 1873; Watson 1896).

Following the Civil War, the specimen collection for the AMM continued with a directive to save specimens from remote forts in the western territories (see Spatola and Pearlstein, this volume). Bentley's response to this directive included collecting and submitting at least three California Indian specimens to the AMM. One skeleton was the remains of Curly-Haired Jack (PS 6287; not to be confused with Captain Jack, the chief of the Modoc and leader during the War) (Otis 1876). Curly-Haired Jack apparently committed suicide on June 8, 1873, with a Colt revolver while in Army custody and was subsequently buried. Bentley, as the battlefield surgeon, presumably knew where Jack was buried and exhumed his freshly buried body, which he then forwarded to the AMM (Jones and Ousley 2008; Riddle 1914).

From there Bentley returned to his post at Point San Jose. He took a leave of absence in the fall of 1874 to serve as the first superintendent of the Napa Insane Asylum. While there, he supervised the construction of the hospital and studied mental diseases and the pathological tissues and structure of the nervous system (California State Legislature [CSL] 1876; White & Co. 1896). During the same period, he served as a member of the Army Retiring Board and recorder of the medical examining board (Watson 1896). In 1876, Bentley left Point San Jose to occupy a temporary post at Camp Independence, California.

Medical College of the Pacific. While posted at Point San Jose, Bentley served as professor of descriptive and microscopic anatomy and pathology at the Medical College of the Pacific from 1870 to 1875. The school was one of the first medical colleges in San Francisco at a time that the reform of American medical education was becoming an issue of concern to the medical profession (Wilson 1998). The Medical College of the Pacific was later subsumed by Cooper Medical College and ultimately consolidated to become Stanford University School of Medicine. The *Pacific Medical and Surgical Journal* (PMSJ) advertised Bentley's Tuesday lectures on anatomy and special pathology of the respiratory organs as well as Saturday clinics held at the City and County Hospital (PMSJ 1874). Another civil position he held, when not on leave of absence or teaching at the medical college, was as the official pathologist for the City and County Hospital of San Francisco (*The Daily Alta California* 1872a). His work there included performing autopsies of suspicious deaths that occurred in the burgeoning city and testifying in the resultant court cases.

Bentley was elected as delegate and permanent member of the American Medical Association and received accolades from colleagues, such as this published in the PMSJ (1871): "As a practical pathologist, Dr. B. is not excelled, if equaled, on this coast; and we have great confidence in his judgment on the question."

Pathological Deductions. While in California, Bentley penned medical articles published in the *California Medical Gazette*, the *PMSJ*, and the *Boston Medical and Surgical Journal.* The *PMSJ* carried two ongoing series of his: "Pathological Deductions from Conditions Found in the Study of Morbid Anatomy" and "Pathological Conditions Observed in Cases Examined at the City and County Hospital, San Francisco, Cal." The latter series chronicled 80 autopsies from his tenure at the hospital.

California Anatomy Act of 1863–64, amended in 1870. Other chapters will conclude that the human remains at Point San Jose did not represent soldiers (see Cox et al. and Sgheiza and Willey, this volume). The question then arises, who did they represent? That question may be answered by looking at the California Anatomy Act and its provisions for acquisition of cadavers (see also Milligan, this volume). Before the California Anatomy Act passed, the faculty of medical schools in California warned of the extreme measures they would take to access cadavers for medical instruction.

The following is an excerpt from an editorial in the *San Francisco Medical Press* by Elias S. Cooper, founder of the Medical College of the Pacific. He attacks sensational newspaper articles, which chastised the lack of a law against grave robbing. After endorsing the act of body snatching, Cooper (1860:242) continues:

> But they cannot succeed in breaking up the dissections. If necessary, those concerned will send all over the State or coast, for the bodies of criminals who are hanged, or to the Indian Reservations, even if each cadaver should cost five hundred dollars. The Faculty of the Medical Department of the University of the Pacific will recognize no obstacle as too great to be surmounted in the building up a medical school on this coast.

Recognizing the demand for cadavers in Medical training, the California Legislature approved the Anatomy Act in 1864 "to promote the study of Anatomy" (CSL 1864:321). This Act allowed officials (e.g., Coroner or warden of a public hospital) to release the bodies of those who would otherwise be buried at public expense to physicians with the stipulation that the physician post a bond and that after their use, the bodies would be buried in a public cemetery at the physician's expense. Preference for the acquisition of cadavers was given to state-sanctioned medical schools.

The Daily Alta California ([DAC], November 2, 1872, emphasis added) recorded the Commencement Speech by Bentley for the Medical Department of University College (Medical College of the Pacific). In it he extols the benefits of the Anatomy Act in California:

> In the matter of the assistance lent by States to carry on anatomical studies and researches, California took the highest place. There were more liberal enactments in this State for the cultivation of the science than in any other. *The supply of "subjects" was ample for all the requisite needs of the professors and students.* In the Eastern States there was a strong and deeply-rooted prejudice in this regard, which had existed in all times even up to the present day. There, the physician is compelled to risk his life, liberty and reputation to obtain the requisite anatomical knowledge to enable him to intelligently and conscientiously practice his profession. For if detected in obtaining material for dissection, he is denounced as a monster, and is held up to the community as a body snatcher.

Here Bentley references the Anatomy Act and the benefits offered to medical research with the passing of the law. As professor at a medical college, he had the credentials to utilize human cadavers under the provisions of the Anatomy Act. Further, as Pathologist for the City and County Hospital, he had direct access to those most likely to be unable to pay for their own burial.

Little Rock Barracks, Arkansas

After his service in California, Bentley was transferred to Little Rock Barracks, Arkansas, where he served as post surgeon from May 1877 to July 1888. While there, he completed several short, temporary duty assignments elsewhere. He was with troops during the Scranton, Pennsylvania general strike of 1877 in solidarity with the Great Railroad Strike, where Pennsylvania Governor Hartranft declared martial law and had federal and state troops occupy the area. In January of 1878, Bentley was brevetted major surgeon. Following that promotion, Bentley attended to a cholera outbreak in New Orleans, in February 1878. He was on the Texas-Mexico border during the Indian Wars at Fort Clark, Texas, in 1884, as well as Fort Davis and at the southern tip of Texas at Fort Brown in 1886. Between temporary assignments, he returned to Little Rock Barracks, where he was posted until July 3, 1888, when he retired after having reached the requisite age of 64 years old with the rank of lieutenant colonel.

Court-Martial. While in Little Rock as post surgeon, Bentley conducted pathological clinics at his military post. A disgruntled hospital steward was distressed by Bentley's activities, including purported misuse of a cow and some of Bentley's instructional activities. Bentley requested a court of inquiry "to investigate and examine into certain allegations made by Hospital Steward Henry C. Mousell, US Army, affecting his [Assistant Surgeon Bentley's] official conduct" (NARA 1879:1, Exhibit B, 1). The following is an excerpt of the hospital steward's testimony in his letter to the Assistant Adjutant General regarding Bentley's alleged actions (NARA 1879: Exhibit A, 3, emphasis added):

> The Hospital Stewards [*sic*] quarters are in the rear part of the hospital building. Dr. Bentley has a small house which he uses as a dead-house moved to within about ten paces of my door, last winter he got up a school for citizens and *secured the bodies from the City hospital*, had them brought to this dead house. He had about 6 young men

> students, I was with my family compelled to suffer the indignity as well as rudeness of actions which young men of that class will commit, they would come outside of the dead house and in front of my door publicly urinate. The human bodies about six I think when cut up *were buried by Dr. Bentley* scarce twenty paces from my door *in his garden.*

The court, it should be noted, concluded that "nothing has been established as improper" and that no further action was necessary (NARA 1879:39).

Medical Department of Arkansas Industrial University. As in Washington, DC, and San Francisco, Bentley participated in the civic life of Little Rock by serving as a faculty member at the local medical school concurrent with his military duties. When he arrived in the late 1870s, various factions in the medical community were at odds over the organization of a medical school. At issue was the credentialing by local and state organizations. A conference met to reach a consensus and Bentley chaired the meeting (Baird 1979). He was a key proponent and proved essential in convincing the profession and citizens of the need for a medical school recognized by existing regulatory bodies and composed of faculty in good standing with the State Medical Society (Altenberg et al. 1917, Henker 2020). With the establishment of the Medical Department of the Arkansas Industrial University (currently University of Arkansas for Medical Sciences) on October 7, 1879, Bentley gave the opening address and was appointed as the chairs of Surgery and of Theory and Practice of Medicine (Henker 2020).

After the establishment of the medical school, Bentley organized a free clinic, the first of its kind, at the rear of a hardware store in downtown Little Rock (Bair 1979). He was elected president of the Arkansas State Medical Society and the Pulaski County Medical Society, and maintained his relationship with the medical school, serving as dean from 1904 to 1907 (Hempstead 1911). Bentley, one of the most distinguished and well-known physicians in Arkansas, pursued his medical practice until a few weeks before he died at the age of 92 at his home in Little Rock on February 5, 1917. He was buried at Arlington National Cemetery in Alexandria, Virginia.

Conclusion

Reflecting on the Point San Jose pit and its contents, one of the first questions that comes to mind is who might be responsible for the 4,450 human bones encountered behind the post hospital. Bentley's tenure at Howard

University, the Medical College of the Pacific, and the Medical Department of Arkansas Industrial University included classes in pathology and prosection. He also exhibited a proclivity for holding dissection clinics on human cadavers at his military post, and collecting and disposing body parts in dubious locations. He was a top-class surgeon and anatomist who experimented with procedures not in regular practice at the time. These factors coalesce to implicate Bentley as the source of the human remains, a result of benevolent intent to further medical knowledge.

As to the question of where the bodies came from, the issue of access to cadavers was addressed by Bentley himself (see also Milligan, this volume). In his 1872 commencement speech to students of the Medical Department of University College, he extolled the enactment of California's Anatomy Act and the resulting concomitant "ample supply of subjects" (DAC, 1872b). His position as Professor of Descriptive and Microscopic Anatomy and Pathology at the Medical College of the Pacific in San Francisco conferred a preference in the procurement of cadavers. His position as pathologist for the City and County Hospital of San Francisco gave him first-hand access to the recently deceased. Further, he would have had knowledge of the status of those decedents who either did not have the means for a proper burial or did not request one. The testimony of the hospital steward in Little Rock established his source of cadavers during his instruction there as being from the City and County Hospital. The source of the human remains from Point San Jose most likely originated from a population that could not pay for their own burial and who died at the San Francisco City and County Hospital.

Unfortunately, the fires that followed the earthquake of 1906 destroyed countless records of the city of San Francisco. Any certificates from a medical society or bonds to receive bodies in this manner, as required by the Anatomy Act, were accordingly lost. While we can surmise that Bentley's intentions were benevolent in bestowing knowledge to the medical students of San Francisco, if the hypothesis is true, we know that in at least some cases he did not follow through with his obligation to bury properly those studied in a public cemetery. These people who, unbeknownst to them, were utilized for the advancement of science were ultimately cast away in a pit with rubbish from the post hospital.

And so, a surgeon with direct access to cadavers, an interest in anatomy, a proclivity for holding dissection clinics, and burying remains in his garden is undoubtedly the individual most likely responsible for the remains interred at Point San Jose. The initial shock upon encountering a collection of body parts behind a military hospital has offered a fascinating glimpse into late nineteenth-century medical practice and the use of marginalized and lower socioeconomic groups to further medical knowledge.

Acknowledgments

Much credit is due Leo Barker, former Golden Gate National Recreation Area archaeologist, who had previously researched surgeons posted at Point San Jose, the Presidio of San Francisco, and Alcatraz. He proposed Edwin Bentley as a suspect due to Bentley's prolific contributions to the AMM, and interesting history in having served in the Civil War and accompanied soldiers during the Modoc War. I too was intrigued with Bentley's history, but was doubtful, as I was with almost everything Leo said. However, despite my initial reservations, and as on most occasions, in the end, Leo was right. I needed proof to support the hunch though, so began research at the National Archives and Records Administration in Washington, DC, and uncovered evidence indicating Bentley had both the means and background to suggest that he was responsible.

References Cited

Altenberg, C., James Coates, and Charles C. Waters
1917 In Memoriam: Dr. Edwin Bentley: Died February 5, 1917. *Bulletin of the University of Arkansas, Medical Department* 5(4). April, Little Rock.

Baird, W. David
1979 *Medical Education in Arkansas, 1879–1978*. Memphis State University Press, Memphis, Tennessee.

Barnes, Joseph K. (editor)
1870a *The Medical and Surgical History of the War of the Rebellion, 1861–65*. Part I Vol. I. Medical History. Government Printing Office, Washington, DC.
1870b *The Medical and Surgical History of the War of the Rebellion, 1861–65*. Part I Vol. II. Surgical History. Government Printing Office, Washington, DC.
1876 *The Medical and Surgical History of the War of the Rebellion, 1861–65*. Part II Vol. II. Surgical History. Government Printing Office, Washington, DC.
1879 *The Medical and Surgical History of the War of the Rebellion, 1861–65*. Part II Vol. I. Medical History. Government Printing Office, Washington, DC.
1883 *The Medical and Surgical History of the War of the Rebellion, 1861–65*. Part III Vol. II. Surgical History. Government Printing Office, Washington, DC.
1888 *The Medical and Surgical History of the War of the Rebellion, 1861–65*. Part III Vol. I. Medical History. Government Printing Office, Washington, DC.

Battle, Richard
1978 Plastic Surgery in the Two World Wars and in the Years Between. *Journal of the Royal Society of Medicine* 71(11):844–848.

Blaisdell, F. William
1988 Medical Advances during the Civil War: Presidential Address. *Archives of Surgery* 123(9):1045–1050.
Bollett, Alfred Jay
2004 The Major Infectious Epidemic Diseases of Civil War Soldiers. *Infectious Disease Clinics of North America* 18(2):293–309.
Brinton, John H.
1914 *Personal Memoirs of John H. Brinton: Major and Surgeon U.S.V.: 1861–1865.* Neale Publishing, New York.
California State Legislature (CSL)
1864 The Statutes of California, passed at the Fifteenth Session of the Legislature, 1863–64. O. M. Clayes, State Printer, Sacramento.
1876 *Biennial Report of the Board of Directors of the Napa State Asylum for the Insane, for the years 1874 and 1875* in the Appendix to the Journals of the Senate and Assembly of the Twenty-First Session of the Legislature of the State of California. Volume 5. State Printing Office, Sacramento.
Cobb, W. Montague
1980 Edwin Bentley, MD, 1824–1917: Connecticut Yankee in Washington and Little Rock. *Journal of the National Medical Association* 72(10):1012–1014.
Cooper, Elias S.
1860 [Editor's Table: They Cannot Succeed] *San Francisco Medical Press* 1(4):193–256. October, San Francisco.
Cutter, Laura
2015 The Unusual Case of Private George Lemon. *Military Medicine* 180(2):241–242.
Daily Alta California (DAC) [San Francisco, California]
1872a Hospital Pathologist. 31 May. San Francisco, California.
1872b The Commencement Exercises of the Medical Department of University College, Eloquent Addresses by Drs. Edward [*sic*] Bentley, John F. Morse and J. D. Thornton, Esq. 2 November. San Francisco, California.
Guyatt, Mary
2001 Better Legs: Artificial Limbs for British Veterans of the First World War. *Journal of Design History* 14(4):311.
Hammond, William Alexander
1862 *Circular No. 2.* United States Surgeon General's Office, May 1862, Washington City, DC.
Hempstead, Fay
1911 *Historical Review of Arkansas: Its Commerce, Industry and Modern Affairs.* Lewis Publishing, Chicago.
Henker, Fred O.
2020 Edwin Bentley M. D., University of Arkansas for Medical Sciences Founder Biography. https://medicine.uams.edu/about-the-college/college-of-medicine-history/founders/edwin-bentley-m-d/, accessed February 10, 2020.
Hippocrates
1995 *Hippocrates: Volume VIII, Places in Man. Glands. Fleshes. Prorrhetic 1–2. Physician. Use of Liquids. Ulcers. Haemorrhoids and Fistulas* (Loeb Classical Library

No. 482). Edited and translated by Paul Potter. Harvard University Press, Cambridge.

Howard University

1869 *Catalogue of the Officers and Students of Howard University, District of Columbia 1868–69.* Judd and Detweiler, Washington, DC.

Jones, Erica B., and Stephen D. Ousley

2008 The Repatriation Osteology Laboratory, National Museum of Natural History, Smithsonian Institution. In *The Forensic Anthropology Laboratory,* edited by Michael W. Warren, Heather A. Walsh-Haney, and Larel Freas, pp. 118–144. CRC Press, Boca Raton, Florida.

Laffin, J.

1999 *Combat Surgeons.* Sutton, Wiltshire, United Kingdom.

Lamb, Daniel Smith (editor)

1900 Howard University Medical Department: A Historical, Biographical and Statistical Souvenir. *College of Medicine Publications,* Paper 1.

Markland, Billy

2001 Database of US Military Frontier Regular Deaths. Markland and Allied Families Resources. Electronic document, http://freepages.history.rootsweb.ancestry.com/~familyinformation/frontierregs.xls, accessed July 4, 2017.

National Archives and Records Administration (NARA)

1825–1909 Entry No. 92, Point San Jose, California, Summary Data Cards Relating to the Posts and Stations of Medical Officers, 1825–1909. War Department. Office of the Surgeon General. 1818–9/18/1947, Record Group 112: Records of the Office of the Surgeon General (Army), 1775–1994. National Archives Building, Washington, DC.

1806–1902 Entry No. 86, Registers of Military Service of Medical Officers, 1806–1902. Vol. III. War Department. Office of the Surgeon General. 1818–9/18/1947, Record Group 112: Records of the Office of the Surgeon General (Army), 1775–1994. National Archives Building, Washington, DC.

1873 Entry No. 561, Consolidated Military Officer's File of Surgeon Edwin Bentley, Personal Letter from Edwin Bentley to Surgeon General dated May 31st, 1873. Series: Letters Received, 1871–1894, Record Group 94: Records of the Adjutant General's Office, 1762–1984. National Archives, Washington DC.

1879 Bentley, Edwin—Surgeon. Series: Court-Martial Case Files, 12/1800–10/1894, Record Group 153: Records of the Office of the Judge Advocate General (Army), 1792–2010, War Department. Bureau of Military Justice. 1864–1884. National Archives Building, Washington, DC.

1888 Bentley, Edwin—Surgeon. Entry No. 89, Carded Records Relating to Civil War Staff Officers, 1890–1912. Record Group 94: Records of the Adjutant General's Office, 1762–1984. National Archives Building, Washington, DC.

Otis, George A.

1867 *Drawings, Photographs and Lithographs Illustrating the Histories of Seven Survivors of the Operation of Amputation at the Hip-joint, During the War of the Rebellion, Together with Abstracts of these Seven Successful Cases.* United States Surgeon-General's, Office, Washington, DC.

1871 *A Report of Surgical Cases Treated in the Army of the United States from 1865 to 1871,* Circular No. 3, War Department, Surgeon General's Office, August 17, 1871. Government Printing Office, Washington, DC.

1876 *Check List of Preparations and Objects in the Section of Human Anatomy of the United States Army Medical Museum. For Use During the International Exposition of 1876 in Connection with the Representation of the Medical Department of the U. S. Army.* No. 8. Army Medical Museum, Washington, DC.

Pacific Medical and Surgical Journal (PMSJ)

1871 [Editorial] October 1871, 5(53):193–240, San Francisco.

1874 [Advertisement for the Medical College of the Pacific] December 1874, 26 (7):313–364, San Francisco.

Patterson, Robert U.

1934 Personal Letter from Surgeon General Robert U. Patterson, Major General, US Army to Frank Vinsonhaler, dated December 8, 1934.

Powell, William H., and Edward Shippen (editors)

1892 *Officers of the Army and Navy Who Served in the Civil War.* L. R. Hamersly and Company, Philadelphia.

Riddle, Jeff C. Davis

1914 *The Indian History of the Modoc War, and the Causes that Led to It.* Marnell and Company, San Francisco.

Rutkow, Ira M.

2005 *Bleeding Blue and Gray: Civil War Surgery and the Evolution of American Medicine.* Random House, New York.

Sampson, Calvin C.

1967 The Howard Department of Pathology. *Journal of the National Medical Association* 59(6):431–432.

Stansbury, Lynn G., and John R. Hess

2009 Blood Transfusion in World War I: The Roles of Lawrence Bruce Robertson and Oswald Hope Robertson in the "Most Important Medical Advance of the War." *Transfusion Medicine Reviews* 23(3):232–236.

Watson, Irving A. (editor)

1896 *Physicians and Surgeons of America. Illustrated. A Collection of Biographical Sketches of the Regular Medical Profession.* Republican Press Association, Concord, New Hampshire.

Weld, Stanley B.

1965 *Connecticut Physicians in the Civil War.* Connecticut Civil War Centennial Commission, Mansfield Center, Connecticut.

1973 Surgeon Bentley and the First Human Blood Transfusion. *Connecticut Medicine.* 37(8):423–4.

White, James T., and Company

1896 *The National Cyclopaedia of American Biography Being the History of the United States* 6:383. James T. White, New York.

Wilson J.

1998 Stanford University School of Medicine and the Predecessor Schools: A Historical Perspective. Lane Medical Library, Stanford University, Palo Alto, California.

Woodhull, Alfred A.

1866 *Catalogue of the United States Army Medical Museum, Prepared under the Direction of the Surgeon General, U. S. Army.* Government Printing Office, Washington, DC.

II

Archaeology

Following information on the historical background, this second section presents the archaeological context, artifacts, and faunal elements from the Point San Jose pit. As such, it considers the volume's general themes involving the purpose that the discarded remains and artifacts served, what disposal of the bones and artifacts indicated, and what the deposit represented in post–Civil War medicine and military medicine. The materials and their context presented in this section support and guide the interpretations of the human osteological remains in the next section. In doing so, they deepen the conversation concerning the archaeological discovery found at Point San Jose.

5

Discovery, Excavation, and Context of Materials in the Point San Jose Pit

PETER GAVETTE AND P. WILLEY

This chapter provides information on Point San Jose archaeology and the context of materials deposited in the pit. As such, it documents and addresses portions of two guiding questions of the overall work: what purpose had the human remains served and what did their disposal represent? Before delving into those questions, additional background concerning the discovery's comparative context, the specific location's history, and the pit's discovery and excavation are provided.

In North America, most similarly modified skeletal series originated in formally set-aside cemeteries and other sanctified burial places (see Willey et al., this volume; Table 0.1). However, a minority of sites, including Point San Jose, occurred outside formal cemeteries. Those clandestine, non-cemetery contexts provide rich opportunities for understanding the circumstances and purposes of these unorthodox disposals, which the context and the materials' positions elucidate. Unfortunately, discovery of similar clandestine burials often occurred during construction, which compromised the archaeological contexts. Even without that challenge, remarkably few studies have considered spatial analysis of the remains and artifacts. Discoveries at Annapolis, DeSaussure Medical College, and Holden Chapel were among the few studies that have assessed context. The Point San Jose materials are ripe for such a contextual approach, but additional historical background is required before pursuing the materials' placement in the pit.

Historical Overview of Point San Jose and Building 239

The 68.5 acres that comprise the Fort Mason National Historic District are situated on a sand-covered rocky point of land at the northern tip of the San Francisco Peninsula. To the west, extensive wetlands once extended out toward the bay from a barren landscape of shifting sand dunes. The eastern side of the point descends steeply to a small cove and was once covered in native laurels.

Point San Jose served as a Spanish military outpost for nearly three decades, which was subsequently abandoned following Mexican independence. After the signing of the Treaty of Guadalupe Hidalgo, which ceded Alta California to the US in 1848, the US government created a military reservation that encompassed Point San Jose in 1850 (for a broader Bay Area history, see Bright, this volume). In 1864, there were two batteries constructed, one mounted with six 10-inch Rodman cannons, and the other with six 42-pound rifles. After the batteries were built, the headquarters, hospital, and barracks were constructed to accommodate officers and soldiers.

Building 239

Constructed in 1864 as two separate buildings, the 1.5-story Greek revival wood frame structure was originally the post hospital and hospital steward's quarters for Point San Jose (Figure 5.1, top). The eastern building, which was aligned along a north–south axis, was the post hospital and the adjacent east–west oriented building served as the hospital steward's quarters. The two buildings were joined circa 1890 to form a larger L-shaped structure, now known as Building 239. When the buildings were joined, a 12 foot by 16–1/2 foot "dead end" was created in the inside corner of the building (Figure 5.1, bottom).

In an 1875 *Report on the Hygiene of the United States Army* (Billings 1875), the hospital was described by Assistant Surgeons W. A. Bradley, E. J. Marsh, and Edwin Bentley as follows:

> The hospital is a one-story frame building, consisting of a main portion 32½ by 17¼ feet, divided into a ward 16¼ by $13\frac{10}{12}$, a dispensary, 16¼ by $12\frac{10}{12}$ feet in the clear, with a double fire place with a small closet each side between them; adjoining this in the rear is another portion, 19¼ by 17¼ feet, divided into a mess-room, 16¼ by $10\frac{1}{12}$ feet, and a kitchen, 16¼ by $8\frac{1}{6}$ feet, in the clear. The whole building

> is under one hipped roof, and a porch 6 feet wide extends along the front and across the end of the main portion, and the side of addition. The ward is furnished with four iron single bedsteads, giving an air space of 900 cubic feet to each bed; it is well lighted and ventilated. The building is warmed by coal-fires in open grates. A small building near the hospital is occupied for steward's quarters and store-room.
>
> Serious cases of sickness are not treated here. When such occur at the post they are sent for treatment to the hospital at the Presidio. (Billings 1875:517)

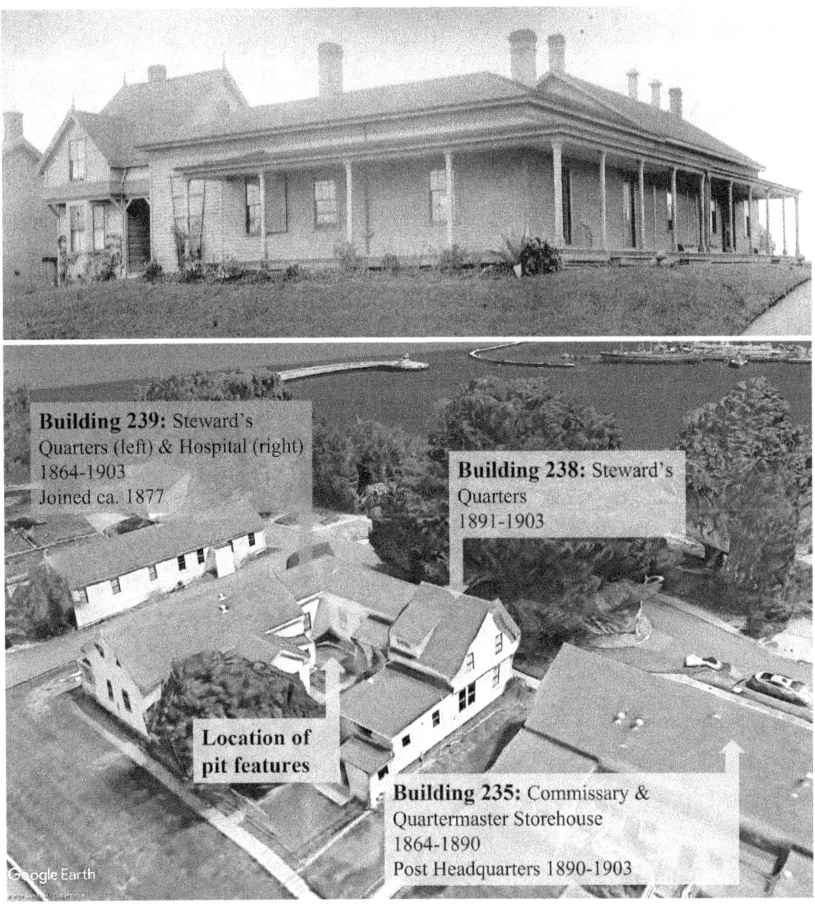

Figure 5.1. *Top*: Point San Jose post hospital (*center*) and steward's quarters (building at left, attached to hospital) facing southwest circa 1891. Courtesy of National Archives, Still Photography Division, Box 10, Series F, (92-F-37-7). *Bottom*: View of hospital and steward's quarters showing alcove location. Image courtesy of Google Earth.

In 1902, Building 201 was constructed to replace the 1864 hospital. Following the hospital's move, Building 239 has served as housing, first as enlisted men's barracks and more recently as market-rate public rental units. Building 239 is currently managed by Golden Gate National Recreation Area (GGNRA), a unit of the National Park Service (NPS).

Rehabilitation of Building 239 and Previous Archaeology

A 2007 *Historic Structure Assessment Report* by a team of seven NPS architects and engineers found Building 239 to "vary from marginally habitable to uninhabitable, due to hidden infrastructure inadequacies (accessibility, structural, mechanical and electrical), hazardous materials, and raccoon infestation" (Cody et al. 2007:1). This structure, listed on the National Register as contributing to the Fort Mason National Historic District (Thompson 1979), was long overdue for comprehensive rehabilitation (Cody et al. 2007).

In 2010 the NPS made plans to rehabilitate Building 239 and other adjacent buildings. In addition to the sensitivity for historical archaeological resources on the grounds of the nineteenth-century military post, two prehistoric sites, with poorly defined boundaries, occurred within the project area. CA-SFR-30, a midden site located west of Building 239 and Pope Road, was discovered during testing in 1978. The test units yielded many diverse artifacts and faunal remains, which suggested a long-term habitation site. Even less was known about the other site, located east of Building 239. The Franklin Street Site (CA-SFR-217), also a midden, was discovered during trenching for a waterline in 2001. Due to potential disturbance in these sensitive areas, the NPS contracted with Holman & Associates in the summer of 2010 to conduct a surface survey and hand augering to identify potential prehistoric and historic-era archaeological deposits in the area. The goals of augering were to determine whether CA-SFR-30 and CA-SFR-217 were a part of the same cultural deposit and if they extended into the area of expected ground disturbance for the project. The surface survey located a few small historical artifacts. The 100 auger borings yielded several historical period artifacts and two chert artifacts, but no intact subsurface midden or artifact-rich deposits were identified (Psota 2010).

The plans to rehabilitate Buildings 235, 238, and 239 included removal of soils with lead contents exceeding the Environmental Protection Agency's regional screening level of 400-ppm for contamination on residential properties. Engineering/Remediation Resources Group, Inc., began the soil remediation in September 2010, focusing on the first couple of feet around the perimeter of the buildings, roughly to the eave drip line. Subsequent testing of the soils beyond the drip line in selected areas indicated lead levels exceeding the 400-ppm limit, so additional removal was conducted in those areas. It was during these additional excavations that the human remains were discovered.

Discovery and Excavation

On Monday afternoon, October 25, 2010, contract archaeologist Tim Spillane made a startling discovery outside Building 239. While monitoring the excavation and remediation of lead-contaminated soils around the perimeter of the building, a cast-iron panel was lifted, revealing human bones. After contacting authorities, the San Francisco Office of the Chief Medical Examiner dispatched Chuck Cecil, a forensic anthropologist, to the site. Cecil determined that it was not a crime scene and relinquished authority and the responsibility of removing the remains to the NPS. The following day, Leo Barker, the GGNRA Park Archaeologist, assembled a crew to begin recovering the rest of the bones and artifacts. The crew consisted of three archaeologists and 18 volunteers, including additional NPS staff members, who participated in the four-day excavation of the site.

The first human bones were found approximately 70 cm below the ground surface covered by a large cast-iron panel. They were located directly below subsequently placed water and gas lines. It is remarkable that these bones remained undisturbed and undiscovered for such a long time considering the installation of those utilities over the years and the 12 auger holes that Holman & Associates dug in 2010 in the small alcove in an attempt to locate such features prior to the remediation work.

Upon discovery, Spillane recovered several long bones. Due to the presence of saw cuts, they were initially thought to be amputations. More long bones were recovered early in the second day of excavations, but they quickly gave way to smaller bones, including a sacrum. That bone was the first indication that this pit did not contain amputated limb bones alone.

Before removing each layer of bones, the skeletal elements were swept free of sand and the area photographed. Commingled with the human remains were glass, ceramic, and metal artifacts. As the bones and artifacts were removed, they were placed in large plastic bags marked with the temporary site name, date, and initials of excavators. The artifacts, which consisted of mostly bottle glass fragments (see Locke Barton, this volume), were bagged separately from the human remains.

Figure 5.2. View of sandstone conglomerate below A Horizon that occurred below adjacent (1891) buildings but absent below Building 239 hospital (1864). Photographs courtesy of National Park Service.

A slightly darker soil staining occurred in the pit, which contrasted with the surrounding matrix. Unfortunately, the homogeneity of the aeolian sands that constitute this area made it difficult to define precisely the edges of the 100 cm by 80 cm oval-shaped intrusive pit. A layer of sandstone conglomerate was identified between the A horizon and the native sand below (Figure 5.2). This layer was placed sometime between 1864 and 1891, because it was found below the foundation of Building 238 (constructed in 1891) but absent beneath the adjacent 1864 buildings. The compacted conglomerate may have been placed either to stabilize the shifting aeolian sands or possibly to cover previous dumping episodes around the hospital.

The first skulls were recovered on the second day of excavation. Excavation on another smaller adjacent pit also started on this date; it was thought to be an extension of the larger pit. The provenience of the artifacts originating from this context was originally referred to as the southern half of the original pit feature. For the most part, other than a small section of four vertebrae in the original pit, no articulations were encountered on this day.

The excavation continued on October 28. More skulls were recovered along with some articulated portions, mostly incomplete sections of vertebral columns. The adjacent smaller pit, approximately 60 cm in diameter, contained mostly artifacts, including complete bottles.

On Friday, October 29, with the weekend and potential for rain approaching, the team resolved to complete excavation. Larger articulations were encountered toward the bottom of the pit, including spinal columns, a few of which included articulated ribs.

The artifacts and human remains were collected and moved to the GGNRA Archaeology Laboratory in the Marin Headlands. Months were spent cleaning, labeling, re-bagging, cataloging, and researching the artifacts. The human remains were similarly brushed clean and re-bagged following the minimal cleaning done in the field. A month after being in the laboratory, the human remains were transferred to large trays, placed in a baker's rack, and ultimately a museum storage cabinet. Approximately 4,450 commingled human remains, 82 faunal remains, and 1,414 artifacts were recovered. The human bones were transferred to the Anthropology Department at California State University, Chico for analysis on April 13, 2016.

Deposit Context

This section considers the context of osseous and artifactual materials in the Point San Jose pit. It assesses placement of the materials in the pit and how those positions reveal behaviors and intentions of the interment. Several questions occur, including:

- When placed in the pit, were human elements treated randomly? Was one element as likely to be placed somewhere in the deposit as any other element? Or did the pit show patterning in the placement of elements?
- Compared to the disarticulated elements, were articulated human segments placed in the pit randomly or was their distribution patterned?

How did the placement of the artifacts (glass, metal, etc.) and faunal bones compare with the human bones distributed in the pit?

Although the homogeneous sandy matrix of the Point San Jose pit obscured the stratigraphy, an understanding of the burial sequence may be inferred based on the depths of the elements and artifacts.

Locations of Bones and Bottles

Based on the principle of superimposition, the lowest materials in the pit (last specimens excavated on October 29, 2010) had been placed in the pit first, and the highest materials had been discarded last (first specimens removed on October 25, 2010). Our approach to the Point San Jose materials embraced this principle employing photographs taken during the excavation.

Reconstruction of stratigraphic and planar bone element and artifact distributions can be established using photographs made by NPS archaeologists during the pit excavation. The archaeologists took 285 photographs during the excavation (dated October 25–29, 2010) and their photographs form the foundation of this assessment.

The photographs showed articulated units, bones vertically and horizontally oriented, skulls, artifacts, and faunal elements. Articulated units consisted of element groups that anatomically connected with one another during life. These articulated units indicated body segments that remained together and intact following death. These units were likely held together by soft tissues and were placed in the pit as segments.

Some excavation photographs depicted vertically oriented limb bones. When set on a flat surface, long bones lie horizontally. But if the pit was too small to accommodate horizontally placed long bones, the bones would instead rest vertically against the sidewalls. Vertical orientation also indicated pit depth as well as side steepness.

Skulls constitute the ultimate anatomical trophy, souvenir, and exemplar (Fabian 2010). This emphasis likely extended to the excavation photographs, and skulls probably were selected for photography as they were uncovered. When the Point San Jose skulls were discarded and placed in the pit, they indicated competing considerations of value, usefulness, and rubbish.

In addition to human bones, artifacts and faunal remains were included in the pit. The excavation photographs were inspected for artifacts and faunal bones for comparison with the distribution of the human remains.

Findings: Field Photographs and Excavation Dates

The earliest deposits (excavated on October 29) suggested a deep pit with steep sides based on element distribution and orientation. At least five limb bones were vertically oriented, indicating the original, steep sides of a deep pit. Some articulated units occurred, including a pelvis with some lumbar vertebrae, at least six right ribs, another nine ribs (possibly right), four left ribs, thoracic vertebrae with left ribs and possibly some right ribs, and ten thoracic vertebrae and some associated ribs. All torsos were placed in supine positions. Another possibly articulated unit consisted of a right radius, ulna, carpals, metacarpals, and phalanges. One skull and a right hemimandible, resting face up, appeared to be articulated. The field photographs on this date showed few—if any—artifacts or faunal elements.

Also in this earliest portion of the deposit, another bone concentration appeared. It included two articulated pelves and a spinal column with articulated ribs. Although the thorax was supine, both pelves were prone. Their position contrasted with the supine position of the other articulated units placed earlier in the pit. This portion of the deposit lacked vertically oriented bones as well as artifacts and faunal elements. These human materials seemed to represent a second pit or perhaps a secondary area within the major pit.

The previous day's (October 28) materials showed broad similarities to the earliest deposits with some key differences. Vertically oriented elements included an articulated tibia and fibula, and femur. Articulated units consisted of the tibia-fibula just mentioned, several vertebra and rib segments, and a pelvis with two lumbar vertebrae. At least one of these groups was supine. Three variously oriented skulls occurred. Two skulls rested on their sides, and one on its top. Some photographs seemed to show sawed articulated ribs. The earliest artifacts deposited appeared in photographs on this day. They consisted of numerous bottles as well as a few buttons associated with scattered human elements. The photographs showed no obvious faunal remains.

Next came the third day (October 27) of excavation. Two elements showed vertical orientation, with the distal ends of the right humerus and femur oriented upward. A few articulated units occurred, including left ribs and possibly associated thoracic vertebrae. The articulated vertebrae exhibited red staining. On the other hand, many scattered elements rested south of the main unit. This day's materials included two skulls; one skull was placed upside down, exposing the cranial base, the other right-side up, exposing its superior surface. One skull had buttons placed in the eye orbits. Broken glass occurred across the deposit.

The second excavation day (October 26) demonstrated both continuity and change compared to the earlier deposits. Only a few ribs rested along the southwest pit wall and showed possible vertical orientation. However, most of the elements consisted of nonarticulated, individual bones. No human skulls or faunal remains appeared in the photographs. On the other hand, many glass and ceramic fragments were evident.

The first excavation date (October 25) represented the culminating final deposit. No limb bones showed vertical orientation, no articulated units occurred, and no skulls appeared in the photographs. Limb bones were placed horizontally and parallel to one another under and beside a cast-iron panel. In addition to the cast-iron panel, glass and ceramics were evident. The only identifiable faunal element from the photographs (a sawed, t-shaped bone) was found on this date.

Based on the excavation images, the sequence of behaviors associated with the pit and disposition can be inferred. The disposal commenced with a decision to discard the remains. A deep, steeply sided pit was dug into sandy deposits with the vertical orientation of some bones in the earlier stages attesting to these assertions. Larger articulated units tended to be placed first, near the bottom of the hole and consisted mostly of ribs and thoraxes. Torsos and pelves tended to be placed in a supine position. After the first deposits, skulls were included, positioned in various, haphazard orientations. Higher in the pit, artifacts, especially glass, were recovered. Continuing up the deposit, fewer limb bones were oriented vertically, indicating a shallower pit. Also, in this later stage, no skulls and fewer articulated units appeared; instead, more separated human elements were scattered among broken glass. In the last stage of burial, no vertically oriented limb bones were found. Some long bones were placed horizontally and parallel to one another; then the deposit was capped with a cast-iron panel. Glass and ceramics appeared most frequently in this last stage of the deposit. Also, the sole faunal element identifiable from the photographs was found near the top. It is possible that perhaps it was introduced later and rested above the pit, not included as part of the original disposal.

Limitations of This Approach

Although much information was extracted concerning context of the materials, certain limitations may impact those conclusions. Limitations included the challenges of using excavation dates as a proxy for burial sequence, photographic selectivity, fragmentation of bones and artifacts, and commingling.

Dates of excavation constituted a crude proxy for the depth below surface of materials as well as the burial sequence. To better understand the disposal process, additional documentation of the context and depth of the pit is required.

Using photographs to establish a sequence of deposition required a systematic, thorough photographic record. If significant periods of the excavation went photographically undocumented or if the photographers emphasized certain elements and features while excluding others, then that selectivity produced gaps in documentation. Such selectivity did occur and is nearly inevitable in photographic documentations. As an example, the human element inventory indicated that tarsals were the most frequently represented human elements (see Cox et al., this volume). None of the excavation photographs, however, showed tarsals. The apparent photographic emphasis on crania has been noted above. Such selective documentation likely distorted the perspectives and interpretations based on the photographs.

Conclusions

After five days of excavation, approximately 4,450 commingled human remains, 82 faunal elements, and more than 1,414 artifacts were recovered from the feature. Bone pits and commingled skeletal series constitute major challenges in their excavation, assessment, documentation, and analysis (e.g., Osterholtz et al. 2013). Excavating and removing the bones from the sand matrix was easy, but the downside of that was the instability of the soil, which caused frequent collapse of the pit walls, making it difficult to both excavate and position oneself adjacent to walls composed of sand. However, even in an ideal situation, striving to collect the intermingled bones as they were deposited in the pit would have proven challenging.

The purposes and intentions behind the Point San Jose disposal pit can be considered along a continuum between two extremes. Were the human remains considered trash, to be dumped as expeditiously as possible? Or, at the other extreme, did the bones' burial represent a notable, terminating event?

Did the Point San Jose pit represent an expeditious effort to discard what constituted trash? Some evidence supported this interpretation. Both waste and human remains were included in the clandestine pit, located far from the closest formal cemetery. Typical of "packing" objects into a confined space, the larger human segments went into the hole first. Then the skulls were deposited in a haphazard manner. Smaller bones and artifacts came last, near the top of the pit. All of these observations suggested that the human remains represented trash and the pit served merely for their convenient disposal.

On the contrary, other evidence indicated a more formal burial, albeit one lacking apparent ritual, unless placement of buttons in the eye orbits of one skull might be considered ritual. First of all, the remains were buried, not deposited with normal post garbage. Trash at Point San Jose was usually discarded in nearby San Francisco Bay (US Surgeon General's Office 1875:517). This written claim was supported by the relative absence of subsurface historical materials (Psota 2010). Most of the thoraxes were placed in the pit in supine positions, not randomly situated. In addition, the pit served foremost as a place for disposal of human bones. Although artifacts were included in the pit, their presence came largely as an afterthought, occurring most frequently near the top of the pit. The final capping of the deposit with the cast-iron panel also suggested a degree of veneration or at least consideration. Although not constituting a headstone, the object's presence indicated respect—perhaps ceremony—beyond merely discarding garbage. It was also possible that the cast-iron panel served as a marker so that the pit location would be permanently recalled.

The pit's spatial context provides insights into the discard and burial events. Now, to return to the general questions and how these materials relate to those driving issues.

The purpose that the human remains served was not as first thought when excavation commenced: a pit for convenient discarding of amputated appendages. The excavation exposed articulated segments and the axial elements, eliminating the amputation explanation. For a more thorough consideration of the remains' purposes, see Peters et al. (this volume).

What did their disposal represent? At one level, they constituted a clandestine burial of remains, ridding those responsible persons of what was considered no longer useful materials. In contrast, some features went beyond mere discard of trash. First of all, the remains were not treated in the manner of standard trash and dumped in the nearby bay. A cast-iron panel capped the deposit suggesting a terminating event and possibly the buttons in the eye orbits of one skull may reflect some respect, perhaps some ceremonial deference.

Despite the circumstances under which the pit was discovered and excavated, we have extracted salient information from the collection of bones and artifacts. The following chapters present and interpret the artifacts and faunal remains from the pit.

References Cited

Billings, J. S.

1875 *Report on the Hygiene of the US Army, with Description of Military Posts.* United States Surgeon General's Office Circular No. 8. Government Printing Office, Washington DC.

Cody, Joanne, Dave Snow, Larry Reynolds, Andy Roberts, Ron Haller, Brian Olson, and Bob Merrick

2007 *Historic Structure Assessment Report for Fort Mason Buildings 235, 238, 240, and 241.* Submitted to National Park Service, Denver Service Center, report on file Golden Gate National Recreation Area.

Fabian, Ann

2010 *The Skull Collectors: Race, Science, and America's Unburied Dead.* ProQuest Ebook Central, University of Chicago Press. https://ebookcentral.proquest.com/lib/csuchico/detail.action?docID=602613, accessed June 9, 2018.

Osterholtz, Anna J., Kathryn M. Baustian, and Debra L. Martin (editors)

2013 *Commingled and Disarticulated Human Remains: Working toward Improved Theory, Method, and Data.* Springer, New York.

Psota, Sunshine

2010 *Archaeological Management Assessment for the Rehabilitation of Buildings 235, 238, and 239 at Fort Mason, Golden Gate National Recreation Area, San Francisco, California.* Submitted to Golden Gate National Recreation Area, National Park Service, San Francisco, On file Golden Gate National Recreation Area.

Thompson, Erwin

1979 "Fort Mason Historic District." National Register of Historic Places Inventory/Nomination Form. Denver Service Center, National Park Service, Denver, April 23, 1979.

6

What Civil War–Era Medicine Bottles Can Reveal about the Point San Jose Pit

ANGELA LOCKE BARTON

The Point San Jose medical waste pit reflects a cross section of life on a mid-nineteenth-century frontier military base. Located in the backyard of the old post hospital, the pit was an undisturbed deposit sealed by a cast-iron panel. The human remains that first signaled this noteworthy find were commingled with artifacts typically associated with this time period (see Gavette and Willey, this volume). Gratifying to any historical archaeologist, this deposit provided an abundance of identifiable and datable artifacts, many of them Civil War–era pharmaceutical bottle fragments. Such bottles are useful in not only establishing relative dating for the deposit, but also reveal the types of medicines used within a military hospital context. The predominance of pharmaceutical bottle fragments reveals an interesting history of developments in bottle-making and medical use spurred on by the tragic realities of the American Civil War as well as the post-war years.

A Hospital Artifact Assemblage Revealed

The artifact assemblage yielded an array of items, indicating activities related to domestic and institutional life. Following a functional classification system analysis (see Blee 1988; Rhodes 1988), the artifacts were organized into four major classes. Domestic artifacts encompass items related to food and beverage, as well as medicines. Personal artifacts are individually owned items, such as clothing, jewelry, and items related to grooming and hygiene. Structural (architectural) artifacts are those associated with the physical presence of a building or structure, including nails and window glass. Activities artifacts include tools and items used in specialized or leisure activities, such as gardening or smoking tobacco. Artifacts like the hair oil bottle and buttons reflect the more personal side of life at the hospital, but the largest category of Point San Jose artifacts is the domestic. This analysis primarily addresses bottles used in the course of medical treatment at the hospital, with a few notable inclusions (see Table 6.1).

Table 6.1. Functional classification of Point San Jose artifacts with counts

Functional Group	Functional Class	Sub-Class	Artifact	Whole	Frags	Min. No.
Domestic	Food Serving	Tableware	Ceramic Whiteware		2	2
	Food Container	Condiment	Lea & Perrin's		39	2
	Bev. Container		Metal Flask	1		1
		Alcohol Bottles	Wine Bottles		27	4
			Tiger Whiskey	1		1
			Ceramic Ale	1	9	4
	Pharm. Supplies	Pharm. Bottles	Hosp. Dept.	1	580	14
			E.R. Squibb		29	7
			H. Thayer & Co.	1		1
			H.P. Wakelee		12	2
			Unidentified	4	672	4
		Pharm. Vials	Pharm. Vials		4	4
		Glass Stoppers	Ground Glass Stoppers	2		2
Personal	Grooming		Hair Oil Bottle	1		1
	Clothing		Prosser Buttons	14		14
			Metal Buttons	2		2
Activities	Tobacco Use		Tobacco Pipe		1	1
	Gardening		Flower Pots		2	2
	Tools		Scissors	1		1
Structural	Window		Window Glass		1	1
	Equipment		Metal Panel		1	1
	Hardware		Lead Washer	1		1
	Light Fixture		Milkglass Frag.		1	1
	Furnishing		Sanitary Porcelain		4	4

One striking feature of the domestic artifacts was the predominance of Civil War–era American manufactured pharmaceutical bottles and bottle fragments. Based on known production dates, most of the identifiable bottle fragments can be dated to a specific timeframe. Another domestic artifact, Lea & Perrin's Worcestershire Sauce bottle fragments, are addressed in this study because this product's bottle characteristics support a relatively tight date range. Almost a hundred faunal bones (see Cole et al., this volume) and two ceramic whiteware sherds are the only other domestic artifacts related to food rather than medicine or alcohol.

Alcohol bottles are included in this analysis because of the fundamental and widespread application of alcohol for medicinal use during this time. Alcoholic beverages, such as whiskey, wine, brandy, and sherry are listed on the Army Medical Supply Table (Surgeon General's Office 1863). Before and during the Civil War, alcohol was used as a general stimulant—for the heart, digestion, and mood—and to treat illnesses such as typhoid and pneumonia (Rothstein 1992:194–195). Alcohol was also the delivery medium for quinine. Although the nature of malaria was not understood until well after the Civil War ended, quinine's effects on the symptoms were known, and it was prescribed frequently to treat fevers of any kind (Rothstein 1992:188).

Despite the fact that it belongs in the personal functional category, rather than domestic, a brief mention will be made of the hair oil bottle because of its dateable manufacturing characteristics, but also to highlight its anomalous presence as a starkly personal item included in a largely institutional deposit.

Civil War–Era Pharmaceutical Bottles

The Army Medical Department began requisitioning standardized bottles for medical supplies when the magnitude of the Civil War became apparent. Civilian pharmaceutical companies had previously met the Army's requirements for supplies, but this system failed with the critical need to meet increasing demand. Surgeons at the front were receiving impure medicines of inconsistent strength. Prices drastically rose due to shortages, market speculation, and drug broker profits. Moreover, the US dollar fell in value when the government's ability to meet financial commitments was questioned (Smith 2001:20).

Shortly after attaining the rank of Surgeon General in 1862, William Alexander Hammond devised a plan to establish laboratories in the city of Philadelphia and the New York City borough of Queens. Money could be saved if the Army processed some of its own medicines and medical supplies. The laboratories could also test the strength and purity of purchased medicines and then repackage products into standard sizes for distribution (Smith 2001:18–22). Hammond established the laboratories under the aegis of the local medical supply depots, so he required no permission or funding approval from Congress. Although criticized for his "improvisation," Hammond ultimately met his goal of providing sufficient quality medicines to the soldiers (Smith 2001: 29–32). Ironically, Hammond was court-martialed in 1864 for "exceeding his legal authority as Surgeon General" (Smith 2001:66).

U.S.A. Hospital Department

Although bottle fragments dominate the Point San Jose pit collection, one intact U.S.A. Hospital Department bottle was recovered. This small bottle is 10.8 cm tall and composed of colorless glass with patent finish (lip) and faint embossed lettering. An additional 580 fragments from a minimum of 13 Hospital Department bottles were identified by the presence of distinctive lettering, unique coloring, and characteristic manufacturing flaws.

The bottle collection contains several fragments from 14 cm tall, oval-shaped, cobalt blue bottles with arched Hospital Department lettering contained within embossed oval panels. By far, the most ubiquitous Hospital Department bottle type represented in the assemblage is the 23.5 cm tall one-liter cylindrical druggist bottle with small-bore and applied double-ring finish. Colors typically vary from amber to olive, as fragments of other colors (cobalt blue, aqua, and colorless) appear to originate from smaller-sized, thinner-walled bottles. Small-bore finishes indicate that these larger bottles probably contained liquid medicines that could be used to mix prescriptions, or perhaps contained other bulk supplies such as chloroform and castor oil, chemicals like turpentine and ammonia, or alcohol (Sternad 2010:49).

From 1862 until 1865, Hospital Department bottles were mouth-blown into post-bottom and key-base molds. They often display distinctive flaws resulting from their crude manufacture: uneven thickness, inclusions of large and small air bubbles and sand, whittle marks, stretch marks, and overspill around applied finishes. The Point San Jose pit specimens exhibit applied double-ring (Figure 6.1, left middle row), prescription, and patent finishes. Cork was the most common closure, but some bottles paired with ground glass stoppers (Dalessandro 2011). Color is inconsistent from bottle to bottle, and sometimes even varies within the same bottle. For example,

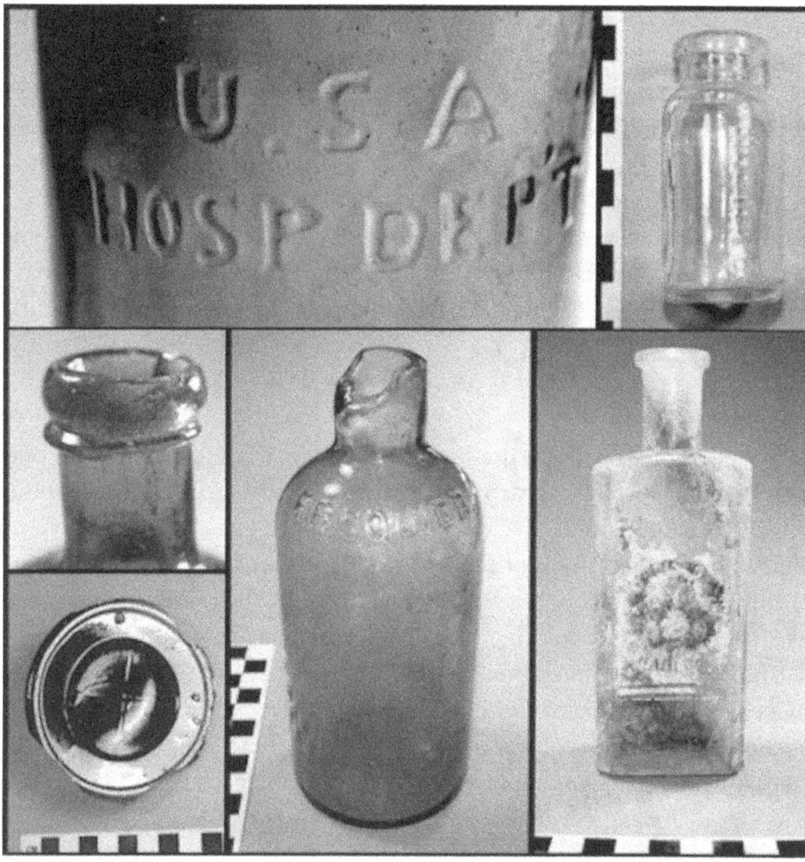

Figure 6.1. Artifacts. *Clockwise from top left*: Large U.S.A. Hospital Department bottle embossed marks; H. Thayer & Co. wide mouth pill bottle; hair oil bottle; E. R. Squibb "druggist" style bottle; Hospital Department base with key mold seams and X mark; Hosp. Dept. bottle neck with crudely applied double ring finish. All photographs by Angela Locke Barton.

olive-colored bottles could be light, medium, dark, smoky, or mixed with amber hues to varying extent. Particular colors and shapes were used for bottles containing poison (Smith 2001:65), however, any specific correlation between color and contents for these bottles has seemingly been lost to history. Paper labels listing the contents of bottles were slightly different for each of the laboratories (Smith 2001:34). Hospital Department bottles have been found in other archaeological contexts with intact applied labels (Dalessandro 2011).

The Hospital Department bottles are embossed on one face: (top row) "U.S.A." and (bottom row) "HOSP. DEPT." (Figure 6.1, top left). The text is arranged in straight lines for most of the bottles, but arched wording is enclosed within oval panels on a few of the smaller bottles. Subtle variation in embossed lettering led one researcher to identify at least 16 different molds (Siri 2009:30–31). Minor differences in appearance are related to place of manufacture. Baltimore Glass Works generally formed bottles with shallow dish-shaped bases and weaker body embossing, whereas bottles produced in Pittsburgh typically had slight push-ups and often had base marks: a raised dot, an asterisk-type star, or the initials "SDS" (Dalessandro 2011). Seven base fragments with no marks from Baltimore exist in the collection. From Pittsburgh, there are three base fragments with asterisks, one fragment with a dot, and four marked with "SDS." Two bases show additional marks not noted in previous studies: a simple line and two crossed lines that form an "X" (Figure 6.1, bottom left).

E. R. Squibb

Among the Point San Jose bottle collection were bottles attributed to Edward R. Squibb, who founded his Brooklyn pharmaceutical company in 1857. While previously employed at the US Naval medical laboratory, Squibb devised a still that safely produced ether of a consistent quality and began supplying that and other pharmaceutical products to the US Army when the Civil War broke out. As the war ramped up, the Medical Department began to produce its own medicines, but Squibb's operation still provided about one-twelfth of the needed supply. In addition, Squibb assisted with organizing the Medical Department laboratories, allowing the full details of his operation to be observed by a government agent (Smith 2001:41).

Identified by the unique letter "Q," 29 recovered fragments represent a minimum of seven Squibb bottles. The bottles exhibit cork seals or have ground interiors for use with glass stoppers. The aqua, teal, and emerald green cylindrical bottles contain the wording: "E. R. SQUIBB." on the shoulder (Figure 6.1, bottom center). The collection's Squibb bottles were all produced before 1892, because in that year Squibb's sons joined the firm and embossing began to include "& SONS" (Echols 1995).

H. Thayer & Co.

Another civilian medical supply provider is represented in the assemblage by an intact H. Thayer & Co. bottle. This colorless 7 cm tall cylindrical bottle is characterized by a wide-bore applied patent finish and maker's mark: "H. THAYER & CO" (from shoulder to heel). The bottle base appears polished, which obscures other manufacturing details. Henry Thayer established his

Cambridge, Massachusetts, laboratory in 1847 and the company is proud of its pharmaceutical contribution to the Civil War effort (Thayers 2019). Thayer specialized in herbal extracts and sugarcoated pills, which was an exciting development in the world of pharmaceutical remedies at the time. The relatively wide-bore dimensions (2 cm) suggest that this bottle contained pills (Figure 6.1, top right). The bottle dates from 1847 to the late 1870s, when most producers of small medicinal bottles had changed from applied to tooled finishes (Lindsey 2010).

H. P. Wakelee

Among the trove of medicinal bottles and fragments comprising the Point San Jose assemblage are fragments from two cylindrical cobalt blue H. P. Wakelee Citrate of Magnesia bottles. Citrate of Magnesia was a popular remedy and laxative that is still in use today. The two base portions indicate that these bottles were mouth-blown into cup-bottom molds. The sole identified Wakelee bottle finish is an applied blob-top. Both of these manufacturing techniques were common for druggist bottles from the 1850s or 1860s continuing into the 1890s (Lindsey 2010). The bottles from the collection read "CITRATE OF MAGNESIA / H. P. WAKELEE / DRUGGIST," and were produced in San Francisco between 1850 and approximately 1875, when the marks began to include "& Co" (Simi 2014). Medical Department policy required purveyors at the medical depots to issue supplies obtained from the Army labs, but allowed purveyors to source some products locally when they could be found at lesser expense (Smith 2001:115).

Unidentified Pharmaceutical/Patent/Folk Remedy Bottles

Numbered among the Point San Jose pit collection were four complete unmarked small pharmaceutical bottles. Also recovered were 672 bottle fragments that appear to be medicinal, based on patent or prescription finishes (lips), oval or rectangular shape, inset panels, thin walls, small size, and narrow bores relative to size (Lindsey 2010). These bottles and fragments lack identifiable maker's marks, but manufacturing details offer production date ranges for some of them. Due to the sheer number of these fragments, many of which had no diagnostic characteristics, a reliable estimate of minimum whole bottles was not attempted.

These pharmaceutical and patent-style bottles might be examples of patent, herbal, and folk remedies not included in the Army's Supply Table. Although Medical Department leadership discouraged their use, many doctors of the time relied upon such remedies (Smith 2001:6). Alternatively, these containers may have been acquired through one of the Army's civilian pharmaceutical providers or been purchased from the private sector. Pharmaceutical bottles of all shapes and sizes were washed and reused at the government labs (Smith 2001:80), which could also account for the presence of non-standard medicinal bottles in the Point San Jose pit. It may have been convenient for the hospital steward to dispense medicines in these small bottles.

Pharmaceutical Vials

The discovery of fragments from four vials adds to the variety of this interesting bottle collection. The vial fragments include one neck with a folded out and ground finish that dates approximately from the 1830s to 1870, as well as a fragment from a 12-sided vial with a pontil mark on the base that indicates likely manufacture before 1870 (Lindsey 2010). Unfortunately, fragments from the other two vials lacked diagnostic manufacturing traits.

Ground Glass Stoppers

One of the exciting finds in the archaeological recovery of the historical bottles are glass bottle stoppers. From the Point San Jose assemblage, two ground glass bottle stoppers were recovered. The first is a colorless stopper with rectangular finial and side mold seams. The second also has a molded rectangular finial and is dark teal. The collection contains two aqua-green E. R. Squibb bottlenecks and four colorless unmarked bottlenecks that are ground on the interior, meant to close with glass stoppers such as these. Large bottles with glass stoppers typically sat on pharmacy ingredient shelves and contained supplies for mixing prescriptions from about the 1840s well into the 1940s (Lindsey 2010).

Dateable Non-Pharmaceutical Bottles

Although they are not strictly pharmaceutical, nearly all of the bottles recovered from the Point San Jose pit could be said to be associated with medical practice. The deposit, clearly institutional in nature, includes artifacts ranging from typical medical treatments to alternative medicines, such as alcohol and Chinese medicine. The condiment bottles and the hair oil bottle are also featured here.

Lea & Perrin's Worcestershire Sauce

A bottle type commonly encountered on nineteenth-century California archaeological sites (Lindsey 2010), 39 fragments from two or more Lea

& Perrin's bottles were recovered from the Point San Jose pit. Worcestershire sauce was a condiment utilized for its ability to "enhance" poorer cuts of meat. These bottles provide a useful date range in association with the medicinal bottles. Light aqua-green body fragments display embossed lettering: "LEA & PERRIN'S" vertically from the shoulder of the bottle to the heel and "WORCESTERSHIRE SAUCE" horizontally around the shoulder. A circular "ACB Co" mark embossed on a base fragment indicates that the bottle was manufactured in England before 1877. After this date, a firm in New York began importing the sauce in barrels and distributing it in American-made bottles (Lunn 1981:10–12). The three-part club sauce finish consists of a small rounded lip, wider flat middle section, and a rounded collar or lower part. A cork-lined glass stopper rested on an interior cork seat ledge to seal this bottle. The crudeness of the finish in conjunction with the base mark signify that these fragments are from imported mouth-blown bottles that date from approximately 1850 to 1877 (Lunn 1981:10–12; Lindsey 2010).

Glass Liquor Bottles

No glass liquor bottles were recovered from the medical waste pit. The Army used its laboratories to test and rebottle medicinal supplies, including alcohol, in standard-sized bottles (Smith 2001:68). It is possible that most liquor supplied to the Point San Jose Army Hospital was packaged in the large Hospital Department bottles (see Sternad 2010:49). Any non-institutional alcohol bottles may have been disposed of by other means.

Glass Wine Bottles

Although no glass liquor bottles are contained within the Point San Jose bottle collection, 27 olive-colored bottle fragments from a minimum of four wine bottles were recovered from the pit. These bottles were identified by their kick ups, or the indented bases characteristic of wine bottles. Three base fragments were recovered from one bottle that was produced in a turn mold, demonstrated by fine horizontal striations along the body, likely during or before the Civil War (Lindsey 2010). The second bottle includes seven fragments from a thick base with kick up and a glass-tipped pontil mark, indicating manufacture before the 1870s (Lindsey 2010). Three related thick light olive fragments comprise the kick up base of a third wine bottle. The remaining 14 fragments are from the kick up base and body of at least one wine bottle, but excessive fragmentation hindered further diagnosis.

Ceramic Ale Bottles

Along with wine bottle fragments, ceramic ale bottles were recovered from the pit. This bottle type is generally ubiquitous in California's late nineteenth-century archaeological record (Lindsey 2010). English companies exported ales and ginger beer (and possibly other beverages) in these durable and often-reused stoneware bottles from the eighteenth into the early twentieth centuries, when the market largely changed over to glass (Wood 2014:104). Few stylistic changes affected these bottles, but the invention of liquid Bristol glaze around 1835 enabled interior glazing and led to their characteristic two-tone beige and brown appearance (Wood 2014:100).

One whole bottle and nine fragments were recovered from the Point San Jose pit. The fragments include one neck and three bases, signifying an additional three bottles. The complete bottle has a grooved ring finish, which secured a wire bale around the cork closure. The neck fragment also has a grooved ring finish and has retained residue from an applied paper label. Of the three base fragments, two have impressed marks on the heel near the base. The first oval-shaped mark reads: "//(UPPER ARCH) PORT DUNDAS / (CENTER) GLASGOW/ (LOWER ARCH) POTTERY CoY." This company operated from 1819 until 1930, and around 1856 converted from salt glaze to a new glaze resulting in a cream-colored appearance (Jewitt 1883:620). The second rectangular partial mark reads: " . . . GRAM(?) / . . . OZ / [GL]ASGOW."

Chinese "Tiger" Whiskey Bottles

Imported bottles support a Chinese presence in mid-nineteenth-century San Francisco. The medical waste pit yielded 19 Chinese Brown Glazed Stoneware fragments that were reassembled to form a complete teardrop-shaped specimen with a constricted neck and a wide (5.5 cm) flared lip. The bottle is 17.5 cm tall and about 12 cm wide at maximum. Based on form, the container likely held medicinal wine or spirits called *ng ka py* or *mei kuei lu* (Choy 2014:15). *Ng ka py* means "wine of five infusions" and is used in traditional Chinese medicine, as well as recreational contexts. *Mei kuei lu* is a type of *baijiu,* a clear liquor that is similar to vodka but distilled from sorghum or rice. Note that although the word *jiu* also translates to "wine," these beverages are 40 to 60 percent alcohol, and would more accurately be classified as spirits (Choy 2014:15). The date range for these bottles is too broad to assist in dating the pit.

Hair Oil Bottle

The lone non-food or non-medicinal bottle recovered was a complete hair oil bottle (see Fig. 6.1 bottom right). This 15 cm tall rectangular bottle (5.6 cm wide, 3.7 cm deep) with chamfered corners has an applied prescription finish, indicating its probable manufacture before the 1890s (Lindsey 2010). One face contains remnants of a paper label depicting flowers and the words "HAIR O. . . ." No brand name or other information is present.

The "Value" of the Point San Jose Bottle Assemblage

Given the variety of the Point San Jose bottle assemblage, the collection serves the historical archaeologist in several valuable ways. First, the bottles provide good relative chronological information in terms of dating the deposit and the former hospital site. Second, the collection also offers a window into specific product data and manufacturing techniques spanning the nineteenth century; these same data were also used to support and supplement dates whenever possible. Finally, the growing number of excavations of hospital sites are slowly adding to the body of literature in this aspect of historical archaeological research. This renders comparative data especially important and helpful in understanding hospital sites like Point San Jose and others.

Chronological Information

Although the Point San Jose bottle collection demonstrates an expansive timeframe—from the early nineteenth century beyond 1940 as shown in Figure 6.2—this timeframe better reflects manufacturing and technological developments rather than the actual date of the deposit as a single event in the archaeological record. However, such developments do provide the opportunity to pin down when the artifacts were disposed of in the pit. For example, the products and technologies represented in the Point San Jose pit collection were in use by 1864 when the hospital was constructed. The Army moved the hospital to a new location in 1902, which serves as a *terminus ante quem* (end date) for the pit. In determining when the artifacts were discarded in the pit, product data and manufacturing techniques can offer some valuable clues.

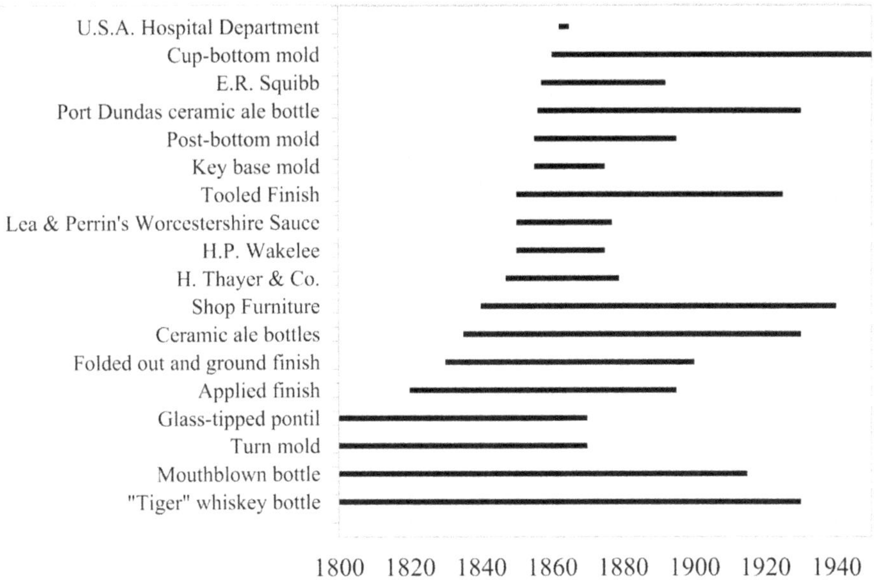

Figure 6.2. Date intervals for product and manufacturing technologies of Point San Jose artifacts.

Product Data

From 1862 until 1865, glass bottle factories at Pittsburgh and Baltimore manufactured standardized U.S.A. Hospital Department bottles in a range of colors and sizes to distribute medicines to the battlefront and frontier military posts. These bottles remained in use past the war's conclusion well into the 1870s, especially at more remote posts (Dalessandro 2011). After the Civil War ended, official wording on the bottles changed from "U.S.A. / HOSP. DEPT." (Figure 6.1) to some version of "U.S.A. / MEDICAL DEPT."

Civil War Army policy required empty pharmaceutical bottles to be returned to the San Francisco medical depot, from where they were shipped to the medical laboratories for reuse. If the bottles were no longer being reused, the money from their sale was to be turned over to the medical purveyor (Surgeon General's Office 1863:7). The Point San Jose bottles were not returned for reuse, nor were they sold at auction, which suggests that pit disposal occurred after these bottles were considered obsolete.

Date ranges from the other bottles in the collection are broader than those from the Hospital Department bottles. The E. R. Squibb bottles were manufactured between 1857 and 1892, when bottle lettering changed to include "& SONS." The H. Thayer & Co. style of bottle was in use from 1847 until the late 1870s. The H. P. Wakelee bottles were manufactured between 1850 and 1875, based on changing bottle marks. The Lea & Perrin's Worcestershire sauce bottles date from 1850 to 1877, as indicated by the base mark and crude finish. The Port Dundas Pottery Company ceramic ale bottle was manufactured between 1856, when the company adopted Bristol glaze, and 1930, when the company closed.

Manufacturing Techniques

When dating a bottle, accuracy increases when detailed product information is available, but the broader dates obtained from manufacturing processes can support a more exact timeline. Several factors influence the precision of these dates. Patent dates indicate when technologies were invented, but producers implemented new machinery and processes at different rates, sometimes leading to "technology lag." The replacement of one technology by a newer one is best indicated by a time range during which both older and newer manufacturing equipment were in use at the same time. Furthermore, new processes were implemented for different types of bottles at different times; for instance, smaller medicinal bottles were consistently produced in cup-bottom molds with tooled finishes before larger bottles (Lindsey 2010).

The glass bottle industry became more refined and increasingly automated during the late nineteenth century. Bottles from the Point San Jose pit are instructive because they represent several manufacturing technologies that were either discontinued or introduced during the 1860s and 1870s. All bottles appear to be mouth-blown into molds, which would have been the norm until about 1900. Bottle molds from the 1820s until about the 1890s did not form the finishes (lips); these would have been hand-formed separately and applied to the bottlenecks, leaving telltale signs: imperfect shape, crude interior (and occasionally exterior) ridges at the site of application, or overspill when liquid glass dripped over the finish (Figure 6.1, left center).

Slightly later tooled finishing was used into the 1920s, and produced a more consistent result, often leaving horizontal striations in the neck area while obscuring the upper portion of the side mold seams. Bottle manufacturers had mostly switched from applied to tooled finishes for smaller pharmaceutical bottles by the late 1870s (Lindsey 2010). The Point San Jose collection includes 43 specimens with applied finishes and four tooled finishes. Most manufacturers adopted automated bottle manufacturing machines between 1910 and 1920. These machines shaped uniform finishes that could take more modern closures, such as the crown cap, which was patented in 1892 (Lindsey 2010). Most of the bottles in the Point San Jose collection closed with a cork and some with a ground glass stopper. No crown finishes exist in the collection.

Bottle bases similarly demonstrate dateable manufacturing technology. Key molds, used from 1855 to the mid-1870s, form bottle bases with two interlocking mold sections designed in such a way to increase bottle stability by eliminating the mold seam in the center of the base (see Figure 6.1, bottom left). Post-bottom molds, used from about 1855 until the 1890s, resulted in a separate circular seam within the resting ring of the bottle (Lindsey 2010). The collection contains 5 fragments from bottles formed in key-base molds and 33 fragments from post-bottom molds. It can be difficult to differentiate seams from these two mold types on incomplete base fragments. Cup-bottom molds were used for small medicine bottles in the 1860s, and slightly later for larger bottles, and are very common today. This mold type eradicates seams along the resting ring by incorporating part of the heel into the base plate. Instead, a horizontal seam rings the side of the bottle just above the base (Lindsey 2010). The collection contains 15 base fragments made in cup-bottom molds.

Two of the four wine bottles in the collection were dated based on manufacturing technologies. The first was formed in a turn mold, as shown by fine horizontal striations along the body. This type of mold was used before and during the Civil War and was largely discontinued afterward. The second wine bottle had a pontil mark on the base, which is a broken lump of glass where a rod held the bottle while it was formed in the mold. Pontils were generally discontinued by the 1870s. The twelve-sided pharmaceutical vial also had a pontil mark, indicating a similar date.

Comparisons with Other Military Hospital Sites

Beyond establishing the date of the deposit, researchers can question why the Point San Jose materials were deposited in the way that they were. Comparing the assemblage to other Civil War–era military hospital collections could provide useful information to that end. Archaeological research at military hospitals is usually part of larger fort-wide projects, but a few studies have focused on hospitals and their artifacts. Hardesty (1981) conducted one of the earliest noteworthy studies at Fort Churchill, Nevada, where written records conflicted with how buildings were used. Consistent with archaeological thinking of the time, Hardesty defined patterns of artifact distribution that could help discern the uses of unknown buildings (Hardesty 1981:288–289). He distinguished six patterns, one of which was the hospital pattern. It could help identify the hospital, and conversely, to predict artifacts likely found at a known hospital site.

Instead of presence or absence of individual artifacts, comparative ratios of types of artifacts in the archaeological record indicate building use (Hardesty 1981:289). The hospital pattern includes a high percentage of spirits artifacts, a medium percentage of structural artifacts (consisting mainly of nails), and relatively low percentages of household artifacts, such as clothing, personal items, and arms (Hardesty 1981:292). Because alcohol was often prescribed medicinally, Hardesty asserts that relatively high percentages of alcohol containers might indicate a post hospital if there is a shortage of other medical artifacts and adequate written records (Hardesty 1981:294).

Trussel (1996) found support for Hardesty's hospital pattern during his 1996 excavations at the Fort Hoskins hospital building in Oregon. Alcohol bottles were recovered predominantly near the hospital, although they also occurred in privies, possibly reflective of "illicit drinking" on a base with a strict no-alcohol policy (Trussel 1996:120–121). Pharmaceutical bottles and medical equipment were scarce at the site due to tight controls and recycling (Trussel 1996:125). He suggests that this dearth might reflect an attempt to sequester contaminated medical waste away from healthy individuals and concludes that a medical refuse dump must be located some distance from the post hospital (Trussel 1996:125–126).

The Bottle Research Group, led by bottle expert Bill Lockhart, analyzed bottle glass found in a Civil War–era hospital privy at Fort Riley, Kansas. Their objective was to infer dates for the excavation strata from dated bottles and fragments. Although the Fort Riley project involved gradual deposition in a privy and the Point San Jose pit was a one-time deposition, the coinciding dates of hospital occupation suggested that there might be similar artifacts found in both collections. Indeed, four examples of Hospital Department bottles were found at Fort Riley, as well as other medicinal bottles and a few Lea & Perrin's Worcestershire Sauce bottles. However, the Fort Riley assemblage contained many glass liquor and beer bottles, which are distinctly lacking in the Fort Mason collection. Most of the diagnostic pharmaceutical bottles found at Fort Riley date from the late 1870s to the 1900s (Lockhart et al. 2012) and probably postdate the Point San Jose medical waste pit deposition.

Contrary to what researchers found at Forts Churchill, Hoskins, and

Riley, very few alcohol bottles were recovered from the Point San Jose pit. If glass alcohol bottles did exist in greater number at Point San Jose, they may have been discarded in the San Francisco Bay where post garbage was customarily thrown (Surgeon General's Office 1875:517). Because an Army medical purveyor was stationed in San Francisco, it is likely that at least some of the medicinal alcohol at Post San Jose was delivered in Hospital Department bottles or other approved containers. If alcohol was received in regular liquor bottles, hospital staff would not have retained the bottles, and they would have been disposed of in the bay.

Comparing the Point San Jose pit artifacts to those found at the other forts indicates that the recovered bottles (and human bones) were not considered "normal" trash, and thus not subject to usual trash disposal practices. The more typical deposits at the other fort hospital sites were amassed over time on the landscape surrounding the buildings or layered in privies. None of the other fort projects involved a one-time pit deposit similar to the Point San Jose pit, although Trussel unsuccessfully searched for a medical waste dump at Fort Hoskins. The foundation trenching that uncovered the Point San Jose pit also unearthed artifacts that had more gradually accumulated along building fringes. It might be fruitful to compare those artifacts to the collections from the other forts.

Conclusion

Using the data from the Point San Jose artifacts, it seems reasonable to assert that the deposit could have occurred during the 1870s. This fits within the general date range of 1864–1902 for the deposition event, and further refines it. Manufacturing details from the pharmaceutical bottles in the collection provide no evidence to refute this hypothesized date; every bottle type found in the pit was in production before or during the Civil War (Figure 6.2). Hospital Department bottles are noted to have been in use into the 1870s at some remote posts (Dalessandro 2011). Because the collection lacks automatically produced bottles or later finish types, such as the crown finish, it is almost certain that the deposit occurred before the turn of the century when these traits became commonplace (see Lindsey 2010). In all probability, these were Civil War–era medicines that had become outdated. Disposing of the Hospital Department bottles in the pit contradicted Army regulations, suggesting that the local medical purveyor also considered the bottles obsolete and indicating that the war had been over for some time.

The assemblage of bottles and bones was treated differently than regular

Point San Jose trash, which was thrown into San Francisco Bay. This difference begs the question, what distinguished the hospital's medical garbage from other post garbage? Burial could indicate a concern with sanitation; perhaps it was thought that the human bones warranted interment rather than disposal. It is intriguing that the pharmaceutical bottles were treated the same as the bones (see Gavette and Willey, this volume). It may have simply seemed expeditious to include the bottles once the pit was dug for the bones.

The creation of the Point San Jose pit could have been related to the transfer of Assistant Surgeon Edwin Bentley away from the post in January of 1875 (see Gavette, this volume). At that time, a shifting medical focus away from the collection of morphologically unusual specimens toward bacteriology (see Spatola and Pearlstein, this volume; see also Rothstein 1992) would have rendered Bentley's anatomical collection obsolete, just like the Civil War–era medicine bottles. The evidence from the artifacts does not indicate what directly precipitated the pit's creation, although the proposed deposit date suggests this coinciding event is a possible candidate.

Acknowledgments

I would like to thank my co-researchers on the project for their many alternative perspectives. Together, these pieces create a fascinating whole. Special thanks to the editors of the volume for their compiling efforts, which included copious feedback and suggestions. Georgia Fox, Anthropology Department Chair at Chico State University, assisted me greatly with final edits. Retired Golden Gate National Recreation Area Archaeologist, Leo Barker obtained funding for my artifact analysis. My work would have been much more difficult without the vastly informative Society for Historical Archaeology's bottle dating website.

References Cited

Blee, Catherine H.

1988 The Moore Cabin and House, Klondike Gold Rush National Historical Park. In *Archeological Investigations in Skagway, Alaska*, Vol. 2. Denver Service Center, National Park Service.

Choy, Philip P.

2014 Interpreting "Overseas Chinese" Ceramics Found on Historical Archaeology Sites: Manufacture, Marks, Classification, and Social Use. Electronic document, https://sha.org/resources/, accessed January 17, 2020.

Dalessandro, Robert J.
2011 US Army Hospital Bottles. Electronic document, http://medicalantiques.com, accessed January 17, 2020.

Echols, Michael
1995 "E. R. Squibb." Electronic document, http://medicalantiques.com, accessed January 17, 2020.

Hardesty, Daniel
1981 Historical Archaeology at Fort Churchill. *Nevada Historical Society Quarterly* 24:283–297.

Jewitt, Llewellynn Frederick William
1883 *The Ceramic Art of Great Britain.* J. S. Virtue, London.

Lindsey, Bill
2010 Bottle Dating. Electronic document, http://www.sha.org/bottle/dating.htm, accessed January 9, 2020.

Lockhart, Bill, Bill Lindsey, Beau Schriever, and Carol Serr
2012 *New Insights from the Bottles Excavated at the Fort Riley Hospital Privy.* Privately published by Bill Lockhart and the Bottle Research Group, Alamogordo, New Mexico.

Lunn, Kevin J.
1981 Identification and Dating of Lea and Perrins' Worcestershire Sauce Bottles on Canadian Historic Sites: Interpretations Past and Present. *Canadian Journal of Archaeology* 5:1–17.

Rhodes, Diane L.
1988 The Mill Creek Dump and the Peniel Mission, Klondike Gold Rush National Historical Park. In *Archeological Investigations in Skagway, Alaska,* Vol. 3. Denver Service Center, National Park Service.

Rothstein, William G.
1992 *American Physicians in the Nineteenth Century: From Sects to Science.* Johns Hopkins University Press, Baltimore. Originally published 1972.

Simi, Rick
2014 "H. P. Wakelee." Electronic document, http://westernbitters.com, accessed January 17, 2020.

Siri, Richard
2009 Quart Hosp. Dept. Molds. *Bottles and Extras* 20(3):30–31.

Smith, George Winston
2001 *Medicines for the Union Army: The United States Army Laboratories during the Civil War.* Haworth Press, Binghamton, New York. Originally published 1962, American Institute of the History of Pharmacy, Madison, Wisconsin.

Sternad, Frank
2010 "U.S.A. Hosp. Dept." *Bottles and Extras* 21(1):48–49.

Thayers [*sic*] Natural Remedies
2019 History. Electronic document, http://thayers.com, accessed January 17, 2020.

Trussel, Timothy
1996 Frontier Military Medicine at Fort Hoskins, 1857–1865: An Archaeological and Historical Perspective. Master's Thesis, Oregon State University, Corvallis.

United States Surgeon General's Office
1863 Circular No. 12 (revised): Directions Concerning the Duties of Medical Purveyors and Medical Storekeepers and the Manner of Obtaining and Accounting for Medical and Hospital Supplies with a Standard Supply Table. Government Printing Office, Washington, DC.
1875 Circular No. 8: A Report on the Hygiene of the United States Army with Descriptions of Military Posts. Government Printing Office, Washington, DC.
Wood, Frank L.
2014 *The World of British Stoneware: It's [sic] History, Manufacture and Wares.* Troubador, Leicester, United Kingdom.

7

Faunal Remains

Food Waste and Animal Dissection at Point San Jose

KASEY E. COLE, KELSIE HART, CECILY MERWIN, AND ALINA TICHININ

Excavations of a pit discovered in historical Point San Jose, California, revealed a commingled assemblage of human and faunal remains and artifacts. This site represents a historical military fort and infirmary. Analysis of the human remains suggests that they may have been used in anatomical dissection studies (see Peters et al., this volume). This chapter describes the nonhuman animal remains recovered from the pit and contributes to our understanding of the context of the assemblage and historical behaviors that occurred at the fort. In historical contexts, studies of faunal remains are used to understand diet and food procurement systems, food discard practices, and studies of ethnicity and social status (see Landon 2005). Accordingly, this study uses the Point San Jose faunal assemblage to evaluate discard practices at the site, gain insight into past dietary behavior, and help interpret the human remains found at the site. The following research questions guide this work:

What species are present in the Point San Jose faunal assemblage?

Does the faunal assemblage show evidence for the anatomical study of animals, or does their presence in the pit result from other activities, such as food waste?

If the Point San Jose faunal assemblage represents food waste, how does the quality of the diet compare to other historical hospitals and military outposts?

Faunal remains recovered from Point San Jose were found mixed with human remains that appear to have been subjected to anatomical study (see Gavette and Willey and Peters et al., this volume). Previous studies document the presence of faunal remains commingled with anatomized human remains (e.g., Boston and Webb 2012; Chapman and Kostro 2017; Fowler and Powers 2012; Hodge 2013; Kausmally 2012; Mann et al. 1991; Rapson et al. 2007). In such assemblages, nonlocal and exotic faunal remains are identified in the assemblages with the presence of longitudinal saw marks and scalpel cut marks that are consistent with dismemberment and the dissection of cadavers for anatomical studies (e.g., Fowler and Powers 2012; Kausmally 2012). Therefore, if the Point San Jose faunal remains resulted from anatomical studies, we expect to find non-food and the possibility of nonlocal taxa in the assemblage and the presence of saw and scalpel cut marks consistent with anatomical studies.

In this study, we also consider that the remains may have been associated with other human activities, such as food consumption and waste. If this is true, then we expect to identify taxa typical of a Euro-American diet during the occupation of Point San Jose, such as domestic cow, pig, and fowl. In addition, we assess the presence of cut marks and portions of bone consistent with historical butchered remains. Previous studies of faunal remains from historical contexts have documented patterns of butchery marks and portions of meat to understand diet quality and food production systems (e.g., Eichelberger 2010; Muñoz 2011; Sanchez 2009; Schulz 1979; Schulz and Gust 1983). These studies used historical accounts of meat prices and butchery descriptions to construct relative rankings of meat cuts based on their economic value. The current analysis considers the relative ranking of meat cuts represented in the Point San Jose faunal assemblage and implications for the quality of diet at the site. Moreover, because the Point San Jose site includes a historic military hospital, a further analysis compares dietary meat cuts with descriptions of military rations and hospital cuisines during late nineteenth-century California. These descriptions were gleaned from historical documents and historical archaeology studies (e.g., Sanchez 2009). These comparisons aid our understanding of food consumption and waste disposal at the site.

Methods

Data Collection

Faunal identifications are made using the Archaeological Laboratory's comparative zooarchaeological collection at California State University, Chico. All unidentified bone fragments were assigned a general size category based on body size. Generally, "small mammal" refers to rodent- or rabbit-sized animals, "medium mammal" indicates coyote- or dog-sized animals, "medium to large mammal" indicates deer- or bighorn sheep-sized animals, and "large mammal" refers to cow- or horse-sized animals. Juvenile specimens were identified based on the observation of unfused epiphyseal surfaces. All modifications to the faunal specimens were recorded. The general condition of each element was categorized as having light, medium, or heavy weathering. Moreover, the presence or absence of root etching, rodent and carnivore gnawing, butchery marks, spiral or horizontal fracture, burning, evidence for bone tool manufacture, and evidence of pathology was identified. Finally, photographs were taken of all skeletal elements that showed butchery, including spiral fractures and saw cut marks. All findings are available in Table B.1, provided online in the University of Florida Digital Collections: https://ufdc.ufl.edu/AA00090501/00012.

Analysis of Cut Marks

We examined each faunal specimen for saw cut marks and incised cut marks. The number of cut marks on each fragment was recorded, as well as the orientation of the cut mark (i.e., transverse or oblique), the tool type (saw or blade), the extent of the cut mark (incomplete or complete), and a description of the anatomical location of the cut mark (see Table B.1 in the UFDC: https://ufdc.ufl.edu/AA00090501/00012). In this analysis, "incomplete" refers to cut marks that primarily affect the cortical surface of the bone, while "complete" indicates cut marks, typically saw cuts, that completely bisect the bone, resulting in fragmentation of the element. The pattern of cut marks represented in the Point San Jose faunal assemblage was then compared to published descriptions of nineteenth-century butchery practices (Muñoz 2011) and animal dissections (Fowler and Powers 2012) to determine whether the remains were from food refuse or were used in comparative anatomy studies. For butchered remains, the anatomical location of cut marks and skeletal portions represented were used to associate faunal remains with specific meat cuts.

Pathology

Although pathology was not the focus of this analysis, the Point San Jose faunal remains were evaluated for apparent signs of trauma and disease as part of the cataloging process. Pathology on the Point San Jose faunal remains was assessed following recommendations from *Standards for Data Collection from Human Skeletal Remains* (Buikstra and Ubelaker 1994; see

Milligan et al., this volume). Evidence for pathology was recorded following three broad categories: shape or size changes, bone loss or formation, and possible disease or trauma. More than one category of pathological change may be present on the same element. Bone loss was further described as either focal or diffuse, and bone formation was further described as either woven (recent, active) or sclerotic (remodeled). Evidence for disease or trauma includes traumatic injuries such as antemortem bone fractures, infection (defined as a bony reaction without traumatic injury), metabolic disease (suggested by porosity on the eye orbits or cranial vault), joint disease and degeneration, and dental disease.

Results

Taxonomic Representation

Ninety-nine faunal specimens are present in the Point San Jose faunal assemblage (Table 7.1). Forty-two specimens were identified to at least the taxonomic level of order, while 57 specimens could not be identified due to weathering and fragmentation.

Avian remains comprise a small fraction of the faunal assemblage. Altogether, four avian specimens were identified. One specimen, a proximal coracoid, was identified to the order Anseriformes (waterfowl), and one proximal left humerus was identified to the family Anatidae (ducks). The remaining two elements were identified as *Gallus gallus* (chicken) and consisted of a complete left femur and a left femur's shaft fragment.

The assemblage is dominated by mammalian remains (n=90). Taphonomic processes limited the identification of many of these remains. Two unidentified elements represent medium mammals, 18 represent medium-large mammals, and 11 are consistent with large mammals. Because of fragmentation, an additional 21 mammalian elements were not categorized by size. Several unidentified mammalian remains compared favorably with domestic pig and cow; however, they lacked diagnostic characteristics to establish a higher level of taxonomic identification.

Of the 38 identified mammalian remains, one element, a right innominate fragment, was identified as *Lepus californicus* (black-tailed jackrabbit). Five remains were identified to the order Carnivora (carnivore). These elements included one complete axis, one anterior atlas that compared favorably with *Canis familiaris* (domestic dog), and three indeterminate cervi-

Table 7.1. Point San Jose faunal taxonomic frequencies and element representation.

Taxa	Mandible	Atlas	Axis	Cervical Vertebra	Thoracic Vertebra	Lumbar Vertebra	Unidentified Vertebra	Rib	Coracoid	Scapula
Aves										
Anseriformes									1	
Anatidae										
Gallus gallus										
Mammalia										
Lepus californicus										
Carnivora		1	1	3						
Artiodactyla								5		
Bos tauras				3	2	4		2		8
Sus scrofa										
Medium Mammal					1			1		
Medium/Large Mammal				1			4	1		3
Large Mammal	1						2	3		
Mammal—Indeterminate								1		
Indeterminate Fragments										
Total	1	1	1	7	3	4	6	13	1	11

(continued)

Table 7.1—*Continued*

Taxa	Humerus	Ulna	Innominate	Illium	Femur	Navicular-cuboid	First Phalanx	Unidentified Longbone	Unidentified Fragment	NISP
Aves										
Anseriformes										1
Anatidae	1									1
Gallus gallus					2					2
Mammalia										
Lepus californicus			1							1
Carnivora										5
Artiodactyla		1		1	1	1	1	1		11
Bos tauras					1					20
Sus scrofa			1							1
Medium Mammal										2
Medium/Large Mammal			1					1	7	18
Large Mammal								2	3	11
Mammal—Indeterminate									20	21
Indeterminate Fragments									5	5
Total	1	1	3	1	4	1	1	4	35	99

cal vertebrae fragments. One cervical vertebra shows evidence of possible butchery.

Thirty-two remains were identified to the order Artiodactyla (even-toed ungulates). Eleven elements could not be identified beyond the level of order. Of these 11, an ilium fragment with cut marks compares favorably with *Ovis aries* (domestic sheep), and a proximal ulna and rib shaft fragment compare favorably with *Capra hircus* (domestic goat). Other elements identified as Artiodactyla include four additional rib fragments, a first phalanx, a navicular-cuboid, and a proximal right femur.

Twenty Artiodactyla specimens were identified as *Bos taurus* (cow). The remains are from the axial skeleton, including cervical vertebrae (n=3), thoracic vertebrae (n=2), lumbar vertebrae (n=4), anterior and posterior rib fragments (n=2), and scapulae (n=8). With the exception of one rib and one thoracic vertebra fragment, all of the remains show evidence of possible butchery. Finally, one innominate fragment was identified as *Sus scrofa* (pig).

Elements Represented

Thirty-six axial elements were identified (Table 7.1). All axial elements are vertebrae or ribs, except for one large mammal mandible fragment. No other cranial elements were identified. The appendicular elements consist of 12 scapula or coracoid fragments, 3 innominate fragments, 10 long bone fragments, and 1 artiodactyl phalanx and tarsal bone. Thirty-five of the 99 specimens in the assemblage could not be identified to element.

Demography

Sixteen juvenile elements are identified based on the presence of unfused epiphyseal surfaces, including two carnivore cervical vertebrae, *Bos taurus* ribs, and cervical, thoracic, and lumbar vertebrae, an artiodactyl rib, femur, and ulna, two mammal vertebrae, and five unidentified fragments of a mammal. Four subadult specimens exhibit signs of possible butchery.

Taphonomy

Most specimens in the faunal assemblage are fragmentary and exhibit light surface weathering. No evidence of rodent gnawing, carnivore damage, or burning was observed. Reddish-brown staining is present on two specimens: one unidentified mammal fragment and one rib fragment from a large mammal (Figure 7.1A). The staining results from an adhered sub-

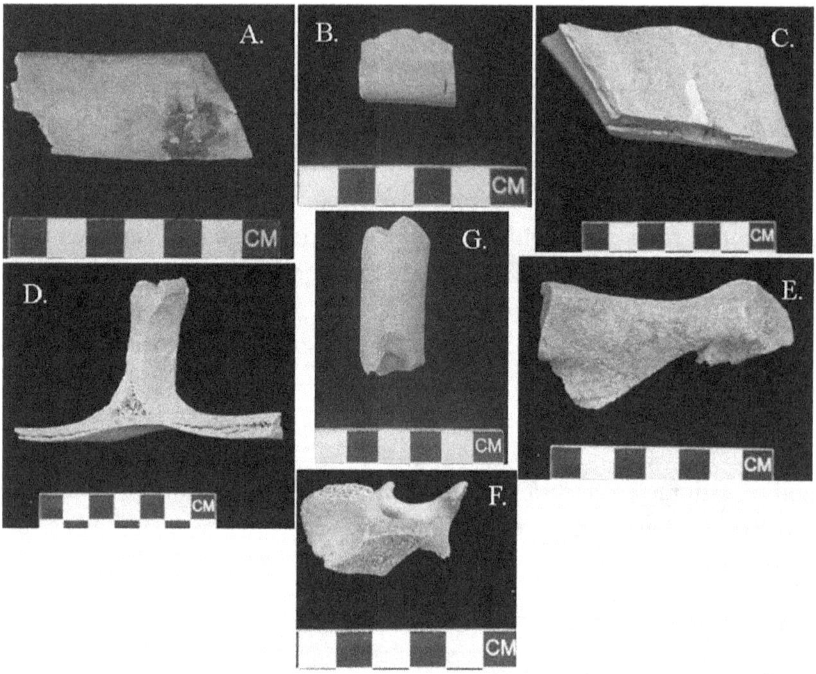

Figure 7.1. Photographs of faunal specimens exhibiting taphonomic modifications. (*A*) Faunal rib shaft fragment (Catalog 43617) showing reddish-brown staining. (*B*) Scapula fragment of a medium- to large-sized mammal (Catalog 43632) with two complete saw cuts on either end of the fragment, plus an incomplete saw cut approximately 2 mm from one of the complete cuts. (*C* and *D*) Scapula fragments of *Bos taurus* (Catalog 43614 and 43637) showing cut marks to the lateral end of the scapula, including the neck and spine. (*E*) Artiodactyla innominate (Catalog 43636) with saw cuts on both ends of the fragment, butchered just above the acetabulum. (*F*) A cervical vertebra of a carnivore (cranial view, Catalog 43605) with a saw cut bisecting the vertebral body. (*G*) Artiodactyl long bone fragment (Catalog 43609) with spiral fractures. Courtesy of National Park Service and Chico State.

stance on the bone surface, possibly rust from close contact with corroding metal.

Evidence of butchery in the form of cut marks is present on 39 specimens representing Carnivora, Artiodactyla, *Bos taurus*, and *Sus scrofa* remains. All of the cut marks are from a toothed instrument (e.g., a saw) rather than incisions from a smooth blade. The majority of butchered bones display only one or two cut marks per fragment. Most of the saw cut marks completely bisect the bone, except for three specimens that show incomplete "false starts" that only penetrate several millimeters into the bone. They consist of an artiodactyl rib, a *Bos taurus* rib, and a scapula fragment of a medium-to-large mammal (compares favorably with *Sus scrofa*) with one false start cut parallel to one of the complete saw cuts (Figure 7.1B).

Seventeen *Bos taurus* specimens display cut marks on the ribs, vertebrae, and scapula. In addition, one unidentified medium-to-large mammal fragment with cut marks compares favorably with the lumbar vertebrae of a cow. One *Bos taurus* rib shows an incomplete saw cut on the margin of the rib shaft. The vertebrae have cut marks in variable locations and orientations. The seven butchered scapula fragments are consistent in the number, location, and orientation of cut marks. Most of the butchered scapula fragments represent transverse sections of bone, bounded by two parallel saw cuts, from the lateral end of the scapula that includes portions of the neck and scapular spine (Figures 7.1C and 7.1D). At least two of the *Bos taurus* fragments represent blade portions of the scapula.

Three medium-to-large mammal scapula fragments compare favorably with a domestic pig. They show the same cut mark patterning as the *Bos taurus* scapula fragments described above.

Additional evidence of butchery is found on Artiodactyla, medium-to-large mammal, and Carnivora remains. One Artiodactyla rib shows four incomplete oblique saw cut marks on its external surface, near the angle. Moreover, two medium-to-large mammal innominate fragments are butchered above the acetabulum (Figure 7.1E). Other medium-to-large mammal fragments that show evidence of butchery include two vertebrae, one unidentified long bone fragment, and one innominate fragment butchered in a similar manner as the Artiodactyla remains described above. Two large mammal ribs and two unidentified large mammal fragments also show cut marks. Finally, one carnivore vertebra shows a complete saw cut through the vertebral body's midline (Figure 7.1F).

Spiral fractures, which occur shortly after death, are present on four fragments. They include an artiodactyl femur, an artiodactyl long bone fragment (Figure 7.1G), and two unidentified large mammal long bone fragments that refit. These fractures may be cultural in origin (e.g., related to marrow extraction), or they may have been caused by natural taphonomic processes (Reitz and Wing 2008:142).

Pathology

Only one specimen exhibits visible signs of pathology. The fragment consists of the rib head, neck, and posterior third of the shaft of a medium-large mammal. The specimen exhibits both bone loss and bone formation,

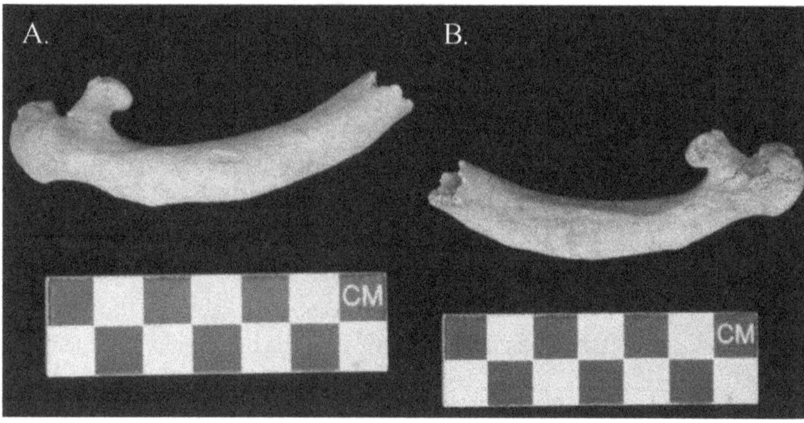

Figure 7.2. Photographs of Point San Jose faunal rib fragment of a medium-large mammal (Catalog 43599) exhibiting pathology. (*A*) Cranial view showing bone formation on the rib shaft and fine porosity on the rib head consistent with bone formation. (*B*) Caudal view showing bone formation and extensive pitting on the rib head and tubercle. Courtesy of National Park Service and Chico State.

resulting in size and shape changes. The rib shaft shows sclerotic (densely remodeled) bone formation on all surfaces, resulting in enlargement of the shaft circumference (Figure 7.2A). Diffuse bone loss is present in the form of fine porosity on the rib head and the caudal surface of the rib shaft. The articular facets of the rib head and tubercle show extensive bone formation and pitting (focal bone loss) (Figure 7.2B). No evidence of trauma was observed on the bone surface; however, a radiographic analysis was not performed to confirm this observation. A differential diagnosis for the inflammation of the rib shaft and joint surfaces was not performed due to the incompleteness of the specimen; however, diagnostic possibilities include a localized infection of the costovertebral joint or a systemic disease process.

Discussion

What Species Are Present in the Point San Jose Faunal Assemblage?

A majority of the taxa were identified as cow and artiodactyl. Some of the remains identified as artiodactyl and medium to large mammals compare favorably with goat and sheep; however, they lack diagnostic characteristics for more specific levels of taxonomic identification. Similarly, many of the remains identified as large mammal compare favorably with a cow, and some of those identified as carnivores compare favorably to a domestic dog.

These results indicated that most of the faunal remains recovered from the pit are likely domesticated medium- or large-size mammals.

Does the Faunal Assemblage Show Evidence for the Anatomical Study of Animals, or Does Their Presence in the Pit Result from Other Activities at the Site, Such As Food Waste?

Cut marks are present on 39 of the 99 faunal specimens, or 39 percent of the assemblage. Cut marks are present on cow, pig, carnivore, artiodactyl, medium-large mammal, and large mammal remains. Elements with cut marks included ribs, vertebrae, scapulae, innominates, and a long bone fragment. All defects are saw cut marks. Most of the saw marks bisect the bone, except for three specimens that show incomplete cuts several millimeters deep. The majority of butchered bones featured only one or two cut marks per fragment.

Anatomical dissection is a possible interpretation for the Point San Jose faunal remains because of their association with anatomized human remains (see Gavette and Willey, this volume). The use of animal remains for vivisection, surgical practice, and comparative anatomy studies are well-documented in the historical and archaeological record (e.g., Boston and Webb 2012; Chapman and Kostro 2017; Fowler and Powers 2012:160; Hodge 2013; Kausmally 2012; Mann et al. 1991; Rapson et al. 2007). These faunal assemblages feature domestic taxa, including cat, rabbit, horse, and fowl, and nonlocal taxa such as monkey, tortoise, hedgehog, fish, frog, and guinea pig. Many of these remains demonstrate evidence of dissection with the same techniques as those used for human remains. Additionally, both nonlocal and domestic taxa have been used in medical teaching. For example, Fowler and Powers (2012) document the use of domestic species for anatomical teaching at the Royal London Hospital between 1820–1854. Here, taphonomic analyses of cow, sheep, goat, and pig, reveal that while many represent food remains, some specimens were indeed dissected. These remains were interpreted by Fowler and Powers (2012) as evidence of dissection, although scalpel cut marks were only found on two specimens. Additionally, the analysis revealed that dog remains predominated the faunal assemblage at the Royal London Hospital. Dogs were also commonly used for anatomical study at the Old Ashmolean Museum and the Craven Street Anatomy School (Fowler and Powers 2012).

Both the butchery and dissection of large animals involve using a saw to separate the carcass into smaller portions. This practice results in transverse and longitudinal saw marks on cervical and lumbar vertebrae, a pattern observed in some specimens in the Point San Jose assemblage. However, most of the saw marks on the Point San Jose faunal remains are consistent with expectations for a nineteenth-century food assemblage. Additionally, many of the Point San Jose faunal remains had evidence of butchery that compared favorably with nineteenth-century meat cuts. Furthermore, unlike the human remains (see Peters et al., this volume), the Point San Jose faunal remains showed no evidence of scalpel cuts associated with excarnation and dissection. Based on the presence of predominately domestic, local taxa that are typical of a Euro-American diet, the absence of scalpel marks, and the presence of saw marks found on elements and in locations typical of butchered meat for consumption, it appears that the faunal remains from the Point San Jose assemblage are consistent with food waste.

If the Point San Jose Faunal Assemblage Represents Food Waste, How Does the Quality of the Diet Compare to Other Historical Hospitals and Military Outposts?

The Point San Jose faunal assemblage is consistent with the "domestic triad" of taxa commonly found in historical contexts: cow, domestic pig, and sheep or goat—all members of the order Artiodactyla (Burke 1994). Avian remains, mainly chicken, are also common in historical domestic settings (Burke 1994).

Previous studies used historical accounts of meat prices and butchery to construct relative rankings of meat cuts based on their economic value (e.g., Bayham et al. 1982; Muñoz 2011; Schulz and Gust 1983). Although secondary meat cuts and their prices have changed considerably over time, historical zooarchaeological studies have demonstrated that primary meat cuts' relative economic ranking remained stable throughout the nineteenth and twentieth centuries (Bayham et al. 1982; Muñoz 2011; Schulz and Gust 1983). The consistency in rankings likely reflects Euro-American perceptions of value based on palatability, tenderness, and fat content (Searls et al. 2005). In general, highly ranked meat cuts came from the middle of the animal, and the lowest-ranked cuts derived from the extremities (Lyman 1979; Schulz 1979).

Table 7.2 summarizes the meat cuts identified in the Point San Jose assemblage and their relative rankings (after Bayham et al. 1982; Schulz and Gust 1983 and Muñoz 2011). Fragments were classified into primary cuts based on their anatomical location. To maximize the data available for in-

Table 7.2. Point San Jose meat cuts identified in the faunal assemblage and their relative rankings by taxa, cut, ranking, and NISP

Taxa	Primary Cut	Relative Ranking[a]	NISP
Cow (*Bos taurus*)			
	Rib	High	1
	Chuck	Medium	8
	Cross/Short Rib	Medium	1
	Sirloin/Short Loin	High	4
	Round		1
	Neck	Low	2
Pig (*Sus scrofa*)			
	Ham	High	1
	Shoulder	Low–Med.	3
C.F. Goat (*Capra hircus*) or Sheep (*Ovis aries*)			
	Sirloin	High	1
	Rib[b]	Med.–High	1
	Foreshank	Low	1
Artiodactyla			
	Rib[b]	Med.–High	2
	Feet[b]	Low	2
	Longbone[b]	N/A	2
Rabbit (*Lepus californicus*)			
	Saddle	Med.–High	1
Fowl—wild (Anatidae)			
	Breast-Upper Wing	N/A	1
Fowl—domestic (*Gallus gallus*)			
	Thigh-Drumstick	N/A	2
Carnivora			
	Neck[b]	N/A	5
Other (Mammal or Unidentified)			
	Head[b]	Low	1
	Various[b]	N/A	41
TOTAL			81

[a] Relative rankings based on Schulz and Gust (1983) and Muñoz (2011).

[b] Anatomical descriptions, not true primary cuts.

terpretation, frequencies in this table included specimens that compared favorably with identified taxa.

Over the last two centuries, beef consumption has been viewed as a status symbol and a sign of material wealth (Sanchez 2009). The Point San Jose assemblage dates to a time in the late nineteenth century when beef was displacing pork as the dominant meat on American tables, which is reflected in the predominance of cow bones at Point San Jose (Horowitz 2006).

Schulz and Gust (1983) tabulated beef prices in nineteenth-century Sacramento to create an ordinal ranking of the economic value of various cuts of beef. Using their ranking, the *Bos taurus* bones from the Point San Jose assemblage represent the following beef cuts: chuck (shoulder area including scapula), neck (cervical vertebrae), rib (proximal ribs and thoracic vertebrae), short rib or cross rib (distal ribs), and short loin (lumbar vertebrae). The distribution of beef cuts in the Point San Jose assemblage is remarkable for its breadth: it includes remains from the costliest portions of the cow, the short loin and ribs; mid-ranked portions of chuck and cross/short rib, representing inexpensive soups, stews, or roasts; and one of the lowest-ranked portions—the neck, which was likely used for soup bones (Schulz and Gust 1983:49).

The butchered pork bones (remains identified as compared favorably with *Sus scrofa)* include portions of the innominate and scapula. These portions represented cuts of ham, a highly ranked cut, and shoulder, a mid-to-low ranked cut (Muñoz 2011:82–83).

Three elements compared favorably with domestic sheep or goat: fragments of ilium, ulna, and rib. These portions corresponded with sirloin, a highly ranked cut, and foreshank, one of the lowest-ranked cuts (Muñoz 2011). The rib fragment could not be assigned to a specific cut due to its lack of diagnostic features, but all of the primary cuts that contain ribs have medium-to-high rankings.

Although the remains identified to the order Artiodactyla could not be attributed to a specific taxon or primary meat cut, some general inferences could be made based on the anatomical location of the fragments. For beef, pork, and goat-sheep (all members of Artiodactyla), primary cuts containing ribs were ranked medium-to-high, while the feet were always one of the lowest-ranked meat cuts (Muñoz 2011). The ranking of long bone fragments varied according to taxon and element, so the artiodactyl long bone fragments were not ranked.

One innominate of a black-tailed jackrabbit was identified in the Point San Jose assemblage. Cottontail and black-tailed jackrabbits were widely available for purchase in late nineteenth-century marketplaces in California and were prepared into various dishes (Gust and Schulz 1983). Based on the relative meat rankings of other mammals, Muñoz (2011) suggested that the middle portions of the rabbit, including the area around the innominate, were more desirable than the head, feet, or limbs.

The Point San Jose assemblage dates to a period of time when there was a shift from the consumption of wildfowl, such as ducks and geese, to the consumption of domestic varieties of chicken and turkey (Simons 1980). This shift was precipitated by the development of large-scale poultry production and the sharp decline of game bird populations in California during the late nineteenth century. Simons (1980) proposes that for sites dating before 1880, a higher ratio of domestic fowl versus wildfowl could indicate greater purchasing power and higher social status because domestic varieties were relatively rare and more expensive at this time. The Point San Jose assemblage is too small to make such an interpretation, but it does contain at least two specimens belonging to *Gallus gallus* and one specimen belonging to the family Anatidae (ducks, geese, and swans).

One of the most surprising findings in the faunal assemblage was the inclusion of five cervical vertebrae belonging to the order Carnivora. The vertebrae did not articulate, and they came from animals of different sizes and ages. The vertebrae lacked diagnostic features for more specific identifications; however, at least one of the vertebrae compared favorably with domestic dog.

One of the cervical vertebrae, belonging to a large carnivore, was bisected by a vertical saw cut through the vertebral body and arch (Figure 7.1D). This cut may have represented butchery; however, the inclusion of carnivore remains is unusual in Euro-American diets. Domestic dog consumption is practiced in many parts of the world and has been documented historically in parts of Asia, Native North America, Africa, and the Pacific Islands (Podberscek 2009). However, the consumption of dogs is generally taboo in Western European cultures, especially in the United Kingdom, and by extension, the US (Podberscek 2009).

The majority of the faunal assemblage could not be identified to a specific taxon, and because the relative ranking of meat cuts is taxon-specific, thus, could not be ranked. One fragment, however, derived from the mandible of a large mammal and compared favorably with horse or cow. Meat from the head of the animal is considered one of the lowest-ranking meat cuts for all taxa, so this fragment was ranked even though it lacked a specific identification.

The interpretation of butchered animal remains from historical contexts is based on the principle that different meat cuts have different economic values, and, therefore, the meat desirability can be reflected in meat cut choices (Huelsbeck 1991; Schulz and Gust 1983). The majority of the Point San Jose faunal assemblage could not be assigned a ranking; however, those ranked meat cuts revealed a relatively large proportion of medium- and high-ranked food sources. Low and low-medium ranked cuts account for only 11.1 percent of the assemblage. These proportions represent the NISP belonging to each category and have not been adjusted for meat yield. Other studies have demonstrated that lower-ranked meat cuts tended to yield less meat than higher-ranked cuts, so the proportions given in Table 7.2 may overestimate the contribution of lower-ranked meats to the diet (Muñoz 2011).

Overall, these results supported the interpretation that the Point San Jose assemblage reflected a medium- to high-quality diet in late nineteenth-century California. Although it was possible that some animals were raised or hunted near Point San Jose, all of the identified taxa (except perhaps the carnivore remains) would have been widely available for purchase in the markets in nearby San Francisco. Furthermore, the few crania and foot bones in the Point San Jose faunal assemblage indicated that butchery activities at the outpost, if any, occurred away from the area of deposition of this assemblage (Boor et al. 2001).

Hospital Diets and Use of Military Rations at Point San Jose

Late nineteenth-century Point San Jose would have been provisioned by the US Army Subsistence Department and the Commissariat Subsistence System to provide standard rations to the men living at the outpost (Risch 1962). The diet likely varied between enlisted men, officers, and infirmary patients. Overall, the Point San Jose faunal assemblage reflects an officers' diet or a typical late nineteenth-century hospital diet with a wide variety of medium- to highly ranked animal cuts. However, the location of the Point San Jose assemblage immediately adjacent to the post infirmary strongly suggests that the faunal assemblage represents hospital food remains.

The Subsistence Department was created by Congressional Act in 1818 and was responsible for the provisioning of Army outposts until 1906, when these duties were reorganized under the Quartermaster Corps (Eichelberger 2010; Hunter 1864). The Commissariat Subsistence System, also established in 1818, coordinated the purchase of bulk goods from public contractors to provision Army outposts (Risch 1962; USWD 1855: par. 1). These goods were transported and distributed by quartermasters to fulfill standard army rations at each outpost.

The Subsistence Department issued enlisted men a standard ration. In 1865, the standard ration included fresh beef, salt beef, salt pork, bacon, bread, beans, rice, peas, coffee, sugar, vinegar, salt, soap, candles, and tobacco (Eichelberger 2010; Risch 1962; USWD 1855; USWD 1861; USWD 1863). Soups or stews served with bread were considered a staple of the soldier's diet. These dishes mixed most of the standard ration items and produced large quantities of food with minimal culinary skills.

Post commanders and soldiers routinely engaged in independent procurement to supplement the standard rations (Eichelberger 2010; Luecke 1990). The Subsistence Department encouraged subsistence farming at Army outposts to make them more self-sufficient, which reduced the demand for supply stores and lowered transportation costs (Break 1975; Luecke 1990; Ball 2001). Individuals also purchased goods from the post sutler's store (a civilian general store licensed to supply a military post) or supplemented their rations through nearby hunting, fishing, or gathering (Eichelberger 2010; Luecke 1990). Dining on wild game was more common for commissioned officers because enlisted men had to be granted furlough or permission to leave the base for these foraging activities while officers did not (Luecke 1990).

In addition to rations, Officers supplemented their own subsistence by purchasing stores from the post Commissary at contract prices or by purchasing goods from the post sutler or nearby merchants (Eichelberger 2010). The higher socioeconomic status of officers was acknowledged through access to additional resources, such as ham, corn meal, dried fruit, coffee, and alcohol, and the ability to purchase higher-quality goods and seasonings (Luecke 1990). Moreover, a major difference between the diets of enlisted men and officers is how the food is prepared. While enlisted men were limited to seasoning provided in the rations, officers were limited by what was available in the post sutler (Luecke 1990). Eichelberger (2010) calculated the total costs of rations for soldiers, officers, volunteer soldiers, and volunteer officers demonstrated that officers consistently consumed a diet that cost more than what enlisted men were issued.

The Point San Jose faunal assemblage includes the remains of domestic taxa, including cow, pig, goat or sheep, and chicken, and wild taxa, including waterfowl and rabbit. The standard ration for enlisted men at Point San Jose was likely restricted to the meat of cows and pigs in the form of fresh beef, salt beef, salt pork, and bacon (Risch 1962; USWD 1855; USWD 1861; USWD 1863). The presence of ham (one *Sus scrofa* innominate with signs of butchery) was consistent with the diet of an enlisted officer (Luecke 1990). Officers would have been able to purchase or hunt foods, such as lamb, mutton, waterfowl, chicken, and rabbit.

Importantly, however, these foods are also consistent with the provisions documented for the post hospital at Fort Yamhill (Eichelberger 2010). At Fort Yamhill, a military outpost in northwest Oregon occupied from 1856–1866, the post hospital procured special rations. Hospital rations included more fresh foods and a much wider variety of foods, such as mutton, lamb, chicken, turkey, venison, eggs, and oysters (Eichelberger 2010). The proportion of medium- and high-quality meat cuts at Point San Jose was also consistent with the findings of Sanchez (2009), who analyzed a sample of faunal remains from Shasta County Hospital, dating from 1855–1900. Although Shasta County Hospital was a public hospital primarily serving the destitute, Sanchez found a large portion of the faunal assemblage consisted of medium- to highly ranked beef cuts.

Conclusion and Summary

The Point San Jose faunal remains represent a food assemblage consistent with an officers' diet or hospital provisions at a late nineteenth-century Army outpost. The location of the assemblage adjacent to the post infirmary strongly suggests that the faunal assemblage represents hospital food remains. The provisions include a variety of medium- to highly ranked meat cuts of beef, pig, and sheep or goat supplemented by wild rabbit, waterfowl, and domestic fowl. The predominance of cow bones in the Point San Jose assemblage is consistent with the rising popularity of beef compared to pork in the American diet, a trend that solidified toward the end of the nineteenth century.

The interpretation of the faunal remains was limited by the small sample size and the paucity of contextual information. A larger sample size, better stratigraphic control, and the addition of archival resources would have facilitated additional insights about foodways at Point San Jose. Furthermore, this analysis may have been biased by the differential identifiability of skeletal fragments (see Grayson 1984). Identifiability is influenced by the number of elements an animal has, differential survivability of various skeletal parts, cultural and natural taphonomic processes, recovery techniques, laboratory procedures, and analytical decisions. Additionally, certain elements and taxa, by nature, are easier to identify.

Finally, the presence of butchered faunal remains (i.e., food waste) and other discards with commingled and disarticulated human skeletal remains in the Point San Jose assemblage highlights the informal disposal of these individuals, which suggests a lower socioeconomic status for these people during life (Nystrom 2017).

Acknowledgments

Thank you to the California State University, Chico, Department of Anthropology for support and advice on this project.

References Cited

Ball, Durwood

2001 *Army Regulars on the Western Frontier.* University of Oklahoma Press. Norman.

Bayham, Frank E., Pamela C. Hatch, and Janet Balsom

1982 *Interpretation of Faunal Remains from the Original Phoenix Townsite, Blocks 1 and 2.* Department of Anthropology, Arizona State University, Tempe, Arizona.

Boor, Jocelyn, Saul Drake, and Vanesa Zietz

2001 A Preliminary Analysis of the Faunal Remains from the Trimborn Farm Site. In *Program in Midwestern Archaeology (Southeastern Wisconsin Archaeology Program): 2000–2001.* Report of Investigations No. 148, edited by Robert J. Jeske, pp. 124–140. Historic Preservation Division Wisconsin Historical Society, Madison.

Boston, Ceridwen, and Helen Webb

2012 Early Medical Training and Treatment in Oxford: A Consideration of the Archaeological and Historical Evidence. In *Anatomical Dissection and Enlightenment England and Beyond: Autopsy, Pathology, and Display,* edited by Piers Mitchell, pp. 43–68. Ashgate, Burlington, New Jersey.

Burke, Brett

1994 Armory Workers' Domestic Faunal Assemblages. In *Domestic Responses to Nineteenth-Century Industrialization: An Archaeology of Park Building 48, Harpers Ferry National Historic Park,* edited by Paul A. Shackel, pp. 7.1–7.24. Occasional Report 12. US Department of the Interior, National Park Service, Washington, DC.

Chapman, Ellen, and Mark Kostro

2017 A Dissection at the Coffeehouse? The Performance of Anatomical Expertise in Colonial America. In *The Bioarchaeology of Dissection and Autopsy in the United States'* edited by Kenneth C. Nystrom, pp. 61–76. Springer International Publishing, Heidelberg.

Eichelberger, Justin E.
2010 *Foodways at Fort Yamhill, 1856–1866: An Archaeological and Archival Perspective.* Master's thesis, Anthropology Department, Oregon State University, Corvallis.
Fowler, Louise, and Natasha Powers
2012 *Doctors, Dissection, and Resurrection Men: Excavations in the Nineteenth Century Burial Ground of the London Hospital, 2006.* Museum of London Archaeology Monograph 62. London.
Grayson, Donald K.
1984 *Quantitative Archaeology: Topics in the Analysis of Archaeological Faunas.* Academic Press, Orlando.
Hodge, Christina J.
2013 Non-Bodies of Knowledge: Anatomized Remains from the Holden Chapel Collection, Harvard. *Journal of Social Archaeology* 13(1):122–149.
Horowitz, Roger
2006 *Putting Meat on the American Table: Taste, Technology, Transformation.* The Johns Hopkins University Press, Baltimore.
Huelsbeck, David
1991 Faunal Remains and Consumer Behavior: What is Being Measured? *Historical Archaeology* 25(2):62–76.
Hunter, Capt. R. F.
1864 *Manual for Quartermasters and Commissaries.* D. Van Nostrand, New York.
Kausmally, Tania
2012 William Hewson and the Craven Street Anatomy School. In *Anatomical Dissection in Enlightenment England and Beyond: Autopsy, Pathology, and Display,* edited by Piers Mitchell, pp. 69–76. Ashgate, Burlington, New Jersey.
Landon, David B.
2005 Zooarchaeology and Historical Archaeology: Progress and Prospects. *Journal of Archaeological Method and Theory* 2(1): 1–36.
Luecke, Barbara K.
1990 *Feeding the Frontier Army, 1775–1865.* Grenadier Publications, Eagan.
Lyman, Lee R.
1979 Available Meat from Faunal Remains: A Consideration of Techniques. *American* Antiquity 44(3):536–546.
Mann, Robert W., Douglas W. Owsley, and Paul A. Shackel
1991 A Reconstruction of 19th-Century Surgical Techniques: Bones in Dr. Thompson's Privy. *Historical Archaeology* 25:106–112.
Muñoz, Jennifer Marie
2011 *Faunal Remains as Marker of Ethnic Identity: The Philadelphia House as a Case Study of German-American Ethnicity.* Master's thesis, Department of Anthropology, California State University, Chico.
Nystrom, Kenneth C.
2017 Conclusion. In *The Bioarchaeology of Dissection and Autopsy in the United States,* edited by Kenneth C. Nystrom, pp. 335–344. Springer, New York.

Podberscek, Anthony L.
2009 Good to Pet and Eat: The Keeping and Consuming of Cats and Dogs in South Korea. *Journal of Social Issues* 65(3):615–632.
Rapson, David J., Matthew G. Hill, and George W. Beran
2007 Archaeology of an Early 20th Century Carcass Disposal Pit, Division of Veterinary Medicine, Iowa State College. *Plains Anthropologist* 52(204):373–417.
Reitz, Elizabeth J., and Elizabeth S. Wing
2008 *Zooarchaeology.* Cambridge University Press, New York.
Risch, Erna
1962 *Quartermaster Support of the Army: A History of the Corps, 1775–1939.* Quartermaster Historians Office, Office of the Quartermaster General, Washington.
Sanchez, Rhea Maricar
2009 *Zooarchaeology and Historical Archaeology of Historic Shasta County Hospital 1855–1900: A Case Study.* Master's thesis, Department of Anthropology, California State University, Chico.
Schulz, Peter D.
1979 Historical Faunal Remains from Panamint City: Notes on Diet and Status in a California Boom Town. *Pacific Coast Archaeological Society Quarterly* 15(4):55–63.
Schulz, Peter D., and Sherri M. Gust
1983 Faunal Remains and Social Status in Nineteenth Century Sacramento. *Historical Archaeology* 17(1):44–53.
Searls, G. A., R. J. Maddock, and D. M. Wulf
2005 Intramuscular Tenderness within Four Muscles of the Beef Chuck. *Journal of Animal Science* 83:2835–2842.
Simons, Dwight D.
1980 Bird Remains. In *Historical Archaeology at the Golden Eagle Site,* edited by Mary Praetzellis, Adrian Praetzellis, and Marley R. Brown III, pp. 1-1 to 1-12. Anthropological Studies Center, Sonoma State University, Rohnert Park, California.
United States War Department [USWD]
1855 Regulations of the Subsistence Department, 1855. A.O.P. Nicholson, Washington.
1861 Regulations Concerning Barracks and Quarters for the Army of the United States, 1860. George W. Bowman, Washington.
1863 Regulations for the Subsistence Department of the Army of the United States, 1863. Government Printing Office, Washington.

III

Human Osteology

Following the historical and archaeological sections, this one concerns the human osteology of the Point San Jose pit. It starts with taphonomy, emphasizing incised and saw cuts. Then, human element counts and the number of individuals represented at Point San Jose, as well as attempts to disentangle commingling issues, are presented. An overview of the demography (age, sex, and ancestry) follows. This section continues with general indicators of stress that can be estimated using stature. Isotopic assessments indicate the people's origins and diets. Bone and dental pathology conclude the human osteology section.

These subjects involve many of the general themes that guide our investigation. They include what purposes the discarded human remains had served, what skeletal elements and anatomical segments were selected, the sex, age, ancestry, and migration status of the Point San Jose persons, and indications concerning the peoples' health and stress. The human osteology chapters contribute new understandings of the Point San Jose sample, post–Civil War military medical history, military surgeons, and bioethics. It forms a holistic perspective on the Point San Jose discovery and its implications.

8

Taphonomy of Human Remains from Point San Jose Hospital, San Francisco

MALLORY PETERS, JESSICA CURRY, AND ERIC J. BARTELINK

Taphonomy refers to all processes that affect an organism after death and is defined as the study of the *laws of burial* (Efremov 1940). This spectrum includes both postmortem modifications to remains caused by environmental factors as well as those resulting from human activity. The analysis of postmortem alterations on human remains from Point San Jose provides important contextual information regarding the purpose of the assemblage (e.g., anatomical collection, pathological waste, etc.). This chapter provides an analysis of the taphonomic modifications observed in the Point San Jose assemblage, discusses their potential causes, compares patterns with similar contexts, and addresses the social implications of the burial context.

Skeletal remains can be modified by their environment, by animals, and by people. Therefore, it is critical to be aware of the environmental and social contexts from which human remains are recovered. The pit containing the human remains at Point San Jose was unearthed behind a historical hospital at what is now Fort Mason in San Francisco (Gavette and Willey, this volume). San Francisco has year-round mild temperatures and sandy soils with excellent drainage. The location of the hospital, situated within a heavily populated military fort and nestled at the edge of the water of the bay, would have protected the pit from many types of scavengers that would have otherwise likely dug up the remains. These factors contributed to the postmortem damage identified in the assemblage.

Background

The Point San Jose collection showed a range of different taphonomic alterations, including depositional damage, soil staining, localized staining, root etching, and sharp-force defects. Some of these modifications are the result of the specific depositional environment in which the remains were deposited (e.g., staining, root etching, and postmortem damage), while others reflect human behavior (e.g., sharp-force defects). Factors that contribute to the overall preservation of remains primarily involve the composition and pH of the soil (Baxter 2004), whereas element completeness is based on the same factors as well as portions being lost due to anatomical dissection processes (i.e., what failed to be included in the pit).

Depositional Damage

As discussed earlier (Gavette and Willey, this volume), the remains recovered from Point San Jose were discarded in a pit that was dug into sand rather than first being placed into a burial container, such as a coffin. Thus, the remains were exposed to a variety of environmental conditions and physical forces since their deposition. Because soil chemistry (e.g., pH) and groundwater can weaken the physical and molecular structure of the bones and result in a loss of integrity, these processes can predispose the remains to further damage (Andra-Sorina et al. 2014; Fernández-Jalvo et al. 2010; Kendall et al. 2018; López-Costas et al. 2016). In addition, soil compaction, as well as direct bone-on-bone contact, resulted in further postmortem fragmentation.

Soil Staining

Bones cleaned of soft tissues in a laboratory or allowed to decompose in a protected setting tend to appear off-white or pale yellow in color. However, bones can be stained by soil and other materials, resulting in colors from bright white to various shades of brown and even red or green (Dupras and Schultz 2013). The Point San Jose collection consists of over 4,000 partial and complete skeletal elements (Cox et al., this volume). Soil staining was ubiquitous throughout the collection, and all bones were a similar shade of light brown. The fact that all the elements were interred in the same context (i.e., a sandy pit) resulted in a somewhat uniform exposure of skeletal elements to the same soil and groundwater conditions. Considering the nearly uniform expression of soil staining in the collection, we made no attempt to quantify it.

Localized Staining

Localized staining refers to small, isolated stains visually identifiable on skeletal elements. These stains were easily differentiated from the ubiquitous light brown discoloration found on all elements of the assemblage. While soil staining is caused by being in direct contact with the soil of the depositional environment, localized staining can occur any time prior to, during, or after remains have been placed in the ground. Just as with soil staining, localized staining occurs when the bone is in contact with an object (e.g., rusting metal) for a prolonged period. The chemical reaction between the object and skeletal elements may result in the formation of a localized stain in the area of contact. Additionally, the porous nature of bone makes it easy for materials to stain the exterior surface (Dupras and Schultz 2014).

Root Etching

As with soil staining, root etching is ubiquitous throughout the Point San Jose collection. Decomposing human remains provide resources for plant life, which often results in extensive root systems growing in and around burial sites. When roots come into contact with remains, they excrete acids that break down the bone, allowing uptake of the mineral components by plants from the soil solution (Pokines and Baker 2013). This process leaves an intricate system of meandering linear or curvilinear grooves on the bone's cortex, which can be identified by their U-shaped floor. As with soil staining, root etching was not quantified because it was present on most of the skeletal elements.

Incised Cut Marks and Saw Cut Marks

As mentioned previously, our primary focus was to document the numerous saw cut marks and incised cut marks on the Point San Jose elements and infer their potential causes. Complete saw cut marks that separated the bone were identified by their direction of saw progress and by the distinct striations on kerf walls from the cutting stroke of the saw blade. Incomplete saw cut marks failed to separate a skeletal element and show striations on the kerf walls and a squared-off kerf floor (Symes et al. 2010). In contrast, incised cut marks appear as small, linear defects with V-shaped kerf floors.

Methods

As part of the larger Point San Jose study, we conducted a taphonomic analysis to document postmortem modifications on the skeletal elements and investigate their potential causes (Willey et al., this volume). Depositional damage was identified macroscopically. Elements were examined for evidence of postmortem breaks, crushing damage, and cortical flaking. Soil staining and localized staining were also identified macroscopically, whereas root etching and sharp-force defects were identified using a combination of macroscopic and microscopic visual inspection.

The localized stains were also analyzed using portable X-ray fluorescence (pXRF) to determine whether the stained surfaces demonstrated elevated chemical concentrations that could identify the source of the staining. Portable XRF scans were conducted using a Bruker Tracer IV Series pXRF with S1 PXRF software v. 3.8.30. No filter was utilized for these scans, which allowed us to capture a wider range of chemical elements. A protective plastic screen was placed between the X-ray emitter and the sample material to protect the equipment and ensure that none of the sample material was transferred to the emitter.

All bones with possible evidence of incised cut marks were examined microscopically with a Leica DMS 1000 digital microscope system to differentiate sharp-force defects from root etching. This was done by differentiating the "V" shaped or squared-off kerf floors that result from sharp-force from the "U" shaped kerf floors that are the result of roots dissolving bone to absorb nutrients. Saw mark directionality was determined by identifying the location of breakaway spurs and false start angles by visual inspection under oblique lighting and digital microscopy. Breakaway spurs occur near the terminal end of a saw cut, where the bone fractures before the saw completely bisects the bone. False starts are incomplete saw cut marks where the bone was partially cut and not bisected and are easily identifiable.

Results

The taphonomic analysis focused on assessing postmortem damage and element completeness, root etching, staining patterns, and saw and incised cut marks. Depositional damage and root etching were ubiquitous throughout the collection. General staining was also a common finding, and in most cases, resulted from contact of the remains with soil and groundwater. Additionally, localized bone staining was documented on a few elements,

Table 8.1. Atypical bone staining and pXRF results in a sample from the Point San Jose assemblage

PSJ No.	Element	Stain Description	pXRF Results
41760	Left Tibia	Medium brown to red-brown	Fe, Ar, Kr
41764	Left Tibia	Light brown with bright red	Fe
41778	Left Metatarsal 1	Red	Fe
41781	Left Talus	Medium brown to red	Negative
41974	Right Metacarpal 2	Bright red	Negative
41980	Right Metacarpal 3	Rusty-red	Fe
42040	Right Scapula	Red	Negative
42744	Unsided Hand Phalanx	Bright red	Fe
42923	Proximal Hand Phalanx	Red-orange	Fe
42982	Left Metatarsal 1	Red	Fe
43235	Thoracic Vertebra	Red	Fe
43299	Thoracic Vertebra	Gray	Negative
43441	Cervical Vertebra	Yellow	Negative
43455	Thoracic Vertebra	Yellow	Negative

most notably bright red staining. These modifications were analyzed using pXRF spectrometry to identify the possible sources of the staining (e.g., metal staining vs. organic staining).

Localized staining occurred in only 14 skeletal elements in the Point San Jose collection, including two left tibiae, two left first metatarsals, a left talus, a right second and third metacarpal, a right scapula, two hand phalanges, three thoracic vertebrae, and a cervical vertebra. The stains ranged from red to orange to yellow, with only one instance of gray staining. Of the 14 elements scanned using pXRF (Table 8.1), 6 failed to show elevated concentrations of chemical elements compared to the scanned control surfaces. The remaining eight elements showed elevated concentrations of iron on the stained surface. A likely cause of the red stains is contact with iron-rich oxidizing material in the pit (Dupras and Schultz 2014), although other possibilities cannot be excluded. One left tibia also showed elevated concentrations of arsenic and krypton. The high concentration of arsenic may derive from chemical preservatives used to fix cadaver tissues for dissection (Hodge et al. 2017). The high concentration of krypton was unexpected, and the source of the chemical element is uncertain.

Postmortem modifications, such as saw cut and incised cut marks, were well-represented in the Point San Jose collection, and provide the best

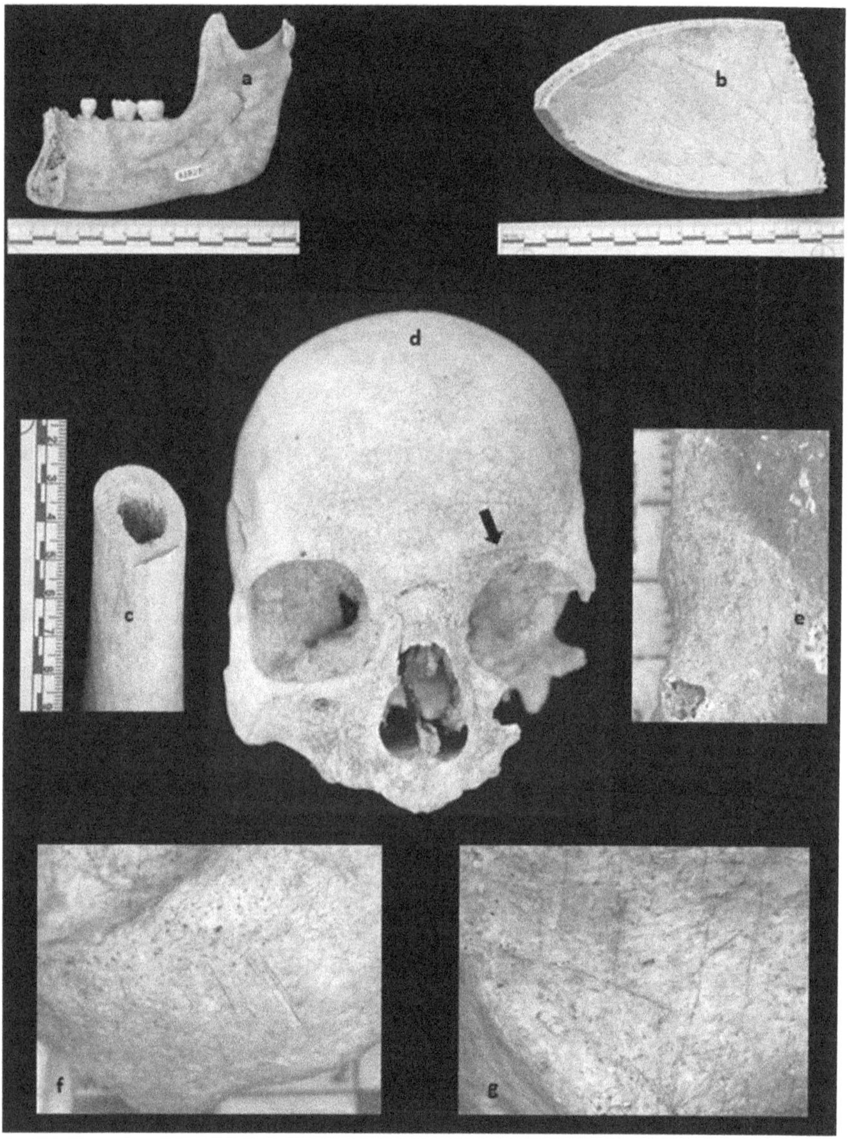

Figure 8.1. (*a*) Bisected right hemi-mandible with evidence of saw cut (42828); (*b*) Left parietal with circumferential and sagittal saw cuts (42908); (*c*) Bisected right femur with saw cut (41770); (*d*) Cranium with incised cut marks superior to left eye orbit (42391); (*e*) Right third metacarpal with red staining (41980); (*f*) Distal right humerus with incised cut marks (42220); and (*g*) Left calcaneus with root etching (41789). Courtesy of National Park Service and Chico State.

evidence for inferring how the collection was used (e.g., autopsy sample, discarded elements from dissection, practice surgery, anatomical teaching collection, etc.) and may help in understanding why it was deposited in the manner that it was (Figure 8.1). Thus, our primary focus was to document and interpret these modifications.

In total, 9.2 percent of the skeletal elements (370/4013) examined demonstrated sharp-force defects, with saw cut marks on 2.1 percent (83/4013) and incised cut marks on 7.2 percent (287/4013) of skeletal elements, respectively (complete and fragmentary elements included). A total of 17 elements showed evidence of both saw cut marks and incised cut marks. These two types of cut marks were distributed throughout all regions of the body, with the highest concentration of saw cut marks observed on sterna and the highest concentration of incised cut marks observed on humeri.

The saw cut mark data are presented by element type and anatomical region in Table 8.2 and Figure 8.2. The highest prevalence of saw cut marks was found on the sternum, followed by the mandible, tibia, cervical verte-

Table 8.2. Distribution of saw and incised cut marks in the Point San Jose sample

	Saw Cut Marks		Incised Cut Marks	
Skeletal Element/Anatomical Region	n/N	%	n/N	%
Cranium	11/192	5.7	5/192	2.6
Mandible	5/41	12.2	5/41	12.2
Sternum	9/33	27.3	1/33	3.0
Clavicle	1/29	3.4	8/29	27.6
Ribs	4/1120	0.4	75/1120	6.7
Cervical	11/131	8.4	25/131	19.1
Thoracic	13/750	1.7	42/750	5.6
Lumbar	8/342	2.3	34/342	9.9
Scapula	0/13	0.0	3/13	23.1
Humerus	1/52	1.9	14/52	26.9
Ulna	2/41	4.9	10/41	24.4
Radius	0/53	0.0	9/53	17.0
Hand	0/414	0.0	13/414	3.1
Pelvis	1/100	1.0	1/100	1.0
Femur	3/50	6.0	12/50	24.0
Tibia	5/53	9.4	2/53	3.8
Fibula	4/51	7.8	6/51	11.8
Foot	5/548	0.7	22/548	4.0
Total	83/4013	2.1	287/4013	7.2

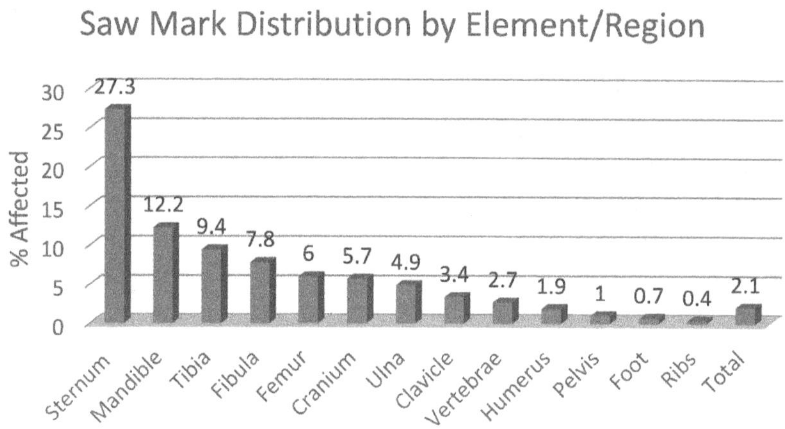

Figure 8.2. Distribution of saw cut marks by skeletal element and anatomical region in the Point San Jose sample.

bra, fibula, femur, cranium, ulna, clavicle, lumbar vertebra, humerus, thoracic vertebra, pelvis, foot, and ribs. No saw marks were identified on the scapula, radius, or hand elements. The data for all vertebrae were combined in Figure 8.2 to group similar elements together.

The incised cut mark data are presented by element type and anatomical region in Table 8.2 and Figure 8.3. The highest prevalence of incised cut marks was found on the clavicle, followed by the humerus, ulna, femur, scapula, cervical vertebra, radius, mandible, fibula, lumbar vertebra, ribs, thoracic vertebrae, foot, tibia, hand, sternum, cranium, and the pelvis. The data for all vertebrae were combined in Figure 8.3 to group similar elements together.

The direction of the cutting stroke for saw cut marks was highly variable and showed little consistency (Table 8.3). Only crania showed evidence of saw cut marks that occurred circumferentially from exterior to interior, consistent with craniotomy to remove the brain (i.e., standard autopsy or dissection procedures). Dittmar and Mitchell (2015) found that many of the skulls from a nineteenth to early twentieth-century human dissection collection from the University of Cambridge, England, surprisingly lacked evidence of craniotomy, but showed external signs of dissection (e.g., incised cut marks). They argued that this suggests retention of intact skulls for curation and teaching. In addition, they argued that skulls with evidence of craniotomy and/or thoracotomy more likely indicate autopsy rather than dissection.

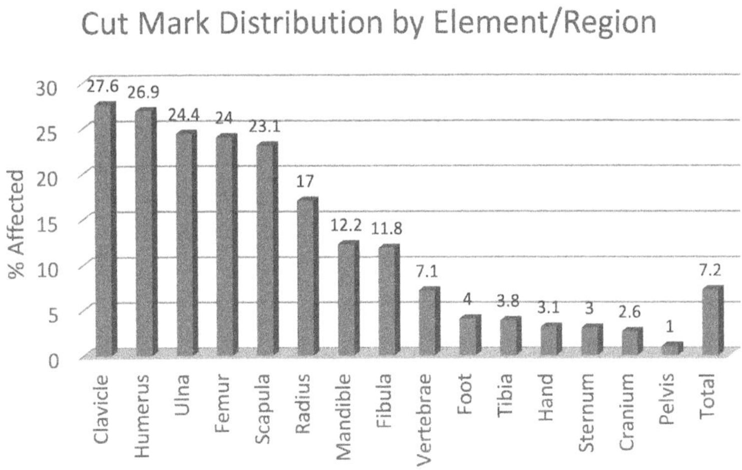

Figure 8.3. Distribution of incised cut marks by skeletal element and anatomical region in the Point San Jose sample.

Table 8.3. Location and direction of saw cut marks in the Point San Jose sample by element

Element	Saw Cut Mark Direction	Location of Defect
Cranium	Exterior to interior	Circumferentially around calvarium
Mandible	Anterior to posterior or lateral to medial	Mandibular symphysis or horizontal ramus
Sternum	Anterior to posterior or posterior to anterior	Midline or through clavicular notch
Vertebrae	Highly variable	Predominantly focused on articular facets
Humerus	Anterior to posterior	Distal midshaft
Ulna	Anterior to posterior	Distal epiphysis
Os Coxa	Inferior to superior	Pubic symphysis
Femur	Posterior to anterior or medial to lateral	Midshaft or femoral neck
Tibia	Highly variable	Epiphyses or distal midshaft
Fibula	Posterior to anterior or medial to lateral	Distal midshaft or distal epihysis
Calcaneus	Superior to inferior	Various locations

Most saw cut marks occurred through easily accessible anatomical regions. For example, the mandibles were sawed either from anterior to posterior or from lateral to medial. Vertebrae were sawed from anterior to posterior, posterior to anterior, or lateral to medial. Sterna were sawed either from anterior to posterior or posterior to anterior. Upper limb bones were sawed from anterior to posterior, although lower limb bones showed variation in direction of saw progress. For example, the femur was sawed from posterior to anterior or medial to lateral, but the tibia and fibula were sawed from medial to lateral, lateral to medial, or posterior to anterior. The degree of variation regarding direction of saw progress within this sample suggests these defects were not created following any kind of standard practice. It is more likely that elements were cut in whatever direction was easiest.

Incised cut marks, generally represented by thin linear cuts, likely indicated dissection and/or removal of soft tissue for inspection of underlying organs, soft tissues, or skeletal elements. These cut marks also varied in their location, likely from the use of scalpel blades to remove soft tissues from bone and to disarticulate skeletal elements. The presence of incised cut marks at areas of articulation, such as the distal end of the humerus, and vertebral bodies and neural arches, suggested that a primary goal was disarticulation.

Discussion

Deposits like the one found at Point San Jose are known from several other contexts within the US. Their existence is probably in large part due to a rising need for medical teaching specimens during a scarcity of resources that relied upon marginalized groups who were not afforded proper burial treatment. This section focuses on the saw and incised cut mark patterns in relation to similar nineteenth-century contexts in the US and contextualizes the Point San Jose assemblage within the broader social structures that created it. For a more in-depth discussion of the social context of cadaver acquisition, see Milligan (this volume) and Hall (this volume).

The distribution of saw and incised cut marks on the skeletal elements from Point San Jose provide important clues regarding the function of the bone assemblage. For example, the high frequency of saw cut marks on sterna (27.3%) reflect opening the chest to expose internal organs, whereas bisected mandibles (12.2%) suggest prosection to expose structures of the skull. However, saw marks were only found on 5.7 percent of crania. Saw marks were more common on lower limb bones (tibia = 9.4%; fibula =

7.8%; and femur = 6.0%) than on upper limb bones (ulna = 4.9%; humerus = 1.9%) and clavicles (3.4%). Other postcranial elements, including the pelvis, vertebrae, ribs, and foot elements were rarely involved (<3.0%), and no saw cuts were identified on scapulae, radii, or hand elements. In addition, the lack of saw cuts on the more intact crania is notable. The distribution of incised cut marks showed a much different pattern than the saw cut mark data, affecting around 25 percent of clavicles, humeri, ulnae, scapulae, and femora. Incised cut marks were less common on radii (17.0%), mandibles (12.2%), fibulae (11.8%), and vertebrae (7.1%), and feet, tibiae, sterna, pelves, and crania were rarely affected (<4.0%).

The distribution of saw cut marks, especially on the postcranial appendicular skeleton, is inconsistent with standard autopsy procedures. Further, saw and incised cut marks are not found in association with healed fractures or other pathological conditions, suggesting that surgical intervention (e.g., amputation, surgery) was not the goal. Instead, these modification patterns are most consistent with anatomized remains (Nystrom 2011, 2014). Anatomized remains may have served as anatomical preparations (e.g., prosected cadavers), teaching specimens, and as experimental models for exploration or surgical training (Owsley 1995).

These findings show some similarities with the dissected remains of at least 12 individuals recovered from a nineteenth-century context at Holden Chapel, Harvard University, where bisected elements included the sternum, femur, and tibia (Hodge et al. 2017:127). Of note, the frequency of saw cut marks on these elements was much higher (50%) than at Point San Jose (<10%). Holden Chapel also showed a high frequency of sectioned ribs, which was not present in the Point San Jose collection, as well as a much higher frequency of incised cuts throughout the skeleton.

At the old Medical College of Georgia, a minimum of 22 anatomized individuals were recovered during renovation of a basement (Harrington and Blakely 1995; Blakely and Harrington 1997). These remains were interred during the mid-nineteenth century when the basement was used as a classroom and laboratory for anatomical dissection by medical students, and the cadavers were likely acquired through grave robbing. In this sample, 40 percent of the elements showed evidence of cut marks (McFarlin and Wineski 1997). The most commonly bisected elements included femora and humeri (5% each), followed by sterna (4.9%), ulnae (4.8%), tibiae (4.6%), radii (3.8%), and clavicles (3.2%) (McFarlin and Wineski 1997:122–123, calculated from Tables 5.1–5.2). The frequencies of bisected elements were lower at the Medical College of Georgia compared to Point

San Jose, with the exception of the ulnae and clavicles, which were similar in both assemblages. Like Holden Chapel, ribs were commonly severed in the Medical College of Georgia, a pattern not identified in the Point San Jose assemblage.

Evidence of anatomical dissection has also been noted in other contexts, including formal cemeteries, dating to the nineteenth century (see Hodge 2012: Table 3 for numerous examples). For example, Novak and Willoughby (2010) reported on evidence of postmortem dissection of two burials from the Spring Street Presbyterian Church in lower Manhattan, dated from 1820–1846. They found evidence of incised cut marks and saw cut marks on the remains of an infant and a child, consistent with removal of the brain for postmortem examination. They suggest these defects likely reflect grave-robbed remains from the cemetery, which were later reinterred following their use as teaching materials. While there is no evidence that links the deposit at Point San Jose to grave robbing, the clandestine deposition of the remains in an undocumented pit suggests a similarly controversial use of human remains.

In another New York context, Nystrom (2014) examined evidence of dissection among 51 individuals interred at the Albany County Almshouse cemetery. Among individuals who had evidence of craniotomy, a few showed small drill holes consistent with rearticulation of the calotte (Lusignan 2004; Solano 2006). This suggests that some of the remains were used as teaching specimens. Transverse cuts were observed through several long bones, and cut marks were observed on vertebrae, a clavicle, and an ischium (Lusignan 2004).

Another historical New York cemetery associated with the Erie County Poorhouse and Buffalo Medical Center showed evidence of dissection of 20 individuals, including craniotomies, transection of long bones, and sectioning of vertebrae (Nystrom 2014). This cemetery appears to demonstrate evidence of both postmortem dissection and probable antemortem surgical amputation. Unlike the pattern at Point San Jose, these three New York cemetery samples show evidence of possible surgical intervention and use of remains as teaching specimens, in addition to evidence of anatomized remains.

One final cemetery project deserves mention, especially given its proximity to Point San Jose. At the nineteenth-century Golden Gate Cemetery (now the Palace of the Legion of Honor), located in San Francisco, roughly 900 burials were disinterred in 1994 (Chatten et al. 1997). While most excavated burials represented intact individuals, there were discoveries of amputated body parts, preserved medical specimens, fetal remains, and medical waste from the cemetery (Buzon et al. 2005; Chatten et al. 1997). The discovery of disposal pits containing embalmed cadavers, a jar containing human eyes, and grave pits containing anatomized remains and commingled amputated limbs indicate that this cemetery was used as a common disposal area for anatomized remains (Chatten et al. 1997:24–33). Much of the original cemetery was used for pauper burials, especially poor, working-class people of Euro-American ancestry, as well as for people of Chinese and Native American ancestry. This cemetery project is relevant to the current project given its proximity and contemporaneity to the Point San Jose hospital. These grave pits may represent the proper and legal disposal of anatomized remains following California state law, whereas Point San Jose represents an improper and illegal disposal event. Unfortunately, little information is available on the results of the Golden Gate Cemetery project (Buzon et al. 2005).

The context of the Point San Jose pit—an undocumented disposal site of at least 25 partial individuals—including animal remains, discarded medical waste, and evidence of dissection, indicates that these individuals were not provided the typical mortuary treatment given to members of society. The incised cut marks reflect dissection for anatomical study, and some of the saw cut marks suggest training (e.g., practice amputations, prosections) by hospital surgeons and their trainees. The saw cut marks show no evidence of life-saving procedures. For example, transversely bisected long bones and mandibles have no evidence of healed trauma or other signs of pathology. Therefore, these elements were likely used as teaching specimens or practice amputations for medical students. Given that the contents of the Point San Jose pit represent a single interment, the remains comprise multiple partial individuals, some elements are underrepresented while others are overrepresented, and animal remains and medical waste are included, it seems probable that the pit represents a clean-up of unwanted dissection materials and other medical detritus. The fact that these people were viewed as objects that could be used, and then disposed of improperly without consequences, is what produced the unique taphonomic modifications observed in the assemblage.

Summary

The remains recovered from the Point San Jose pit showed evidence of general soil staining, localized staining, extensive root etching, depositional damage, and numerous cases of incised and saw cut marks. With the exception of the sharp-force defects and examples of localized staining, all postmortem modifications were caused by the depositional environment. The cut marks observed on the remains resulted from postmortem examination, and indicated use of cadavers for anatomical dissection, and possibly also for practicing medical procedures such as amputations. In contrast to cemetery contexts with evidence of anatomized remains, it is evident that these individuals were not given the burial treatment accorded to them by law for that time period. Instead of being laid to rest with traditional funeral practices, they were buried in secret, conveniently disposed of without the dignity of a proper burial. Structural inequality of the poor and the increasing need for medical specimens resulted in the combination of taphonomic modifications observed in the Point San Jose assemblage.

References Cited

Andra-Sorina, T., O. Ponta, and B. Kelemen

2014 Bone Diagenesis and FTIR Indices: A Correlation. *Studia Universitatis Babes-Bolyai, Biologia* 59:101–113.

Baxter, K.

2004 Extrinsic Factors that Effect the Preservation of Bone. *The Nebraska Anthropologist* 19:38–45.

Blakely, Robert L. and Judith R. Harrington (editors).

1997 *Bones in the Basement: Postmortem Racism in Nineteenth-Century Medical Training.* Smithsonian Institution Press: Washington, DC.

Buzon, Michele R., Phillip L. Walker, Francine D. Verhagen, and Susan L. Kerr

2005 Health and Disease in Nineteenth-Century San Francisco: Skeletal Evidence from a Forgotten Cemetery. *Historical Archaeology* 39:1–15.

Chatten, Cassandra, Katherine Flynn, and Dea Bacchetti

1997 *Excavations at the California Palace of the Legion of Honor.* Report for Abacus Archaeological Associates, Petaluma, California.

Dittmar, J. M., and P. D. Mitchell

2015 A New Method for Identifying and Differentiating Human Dissection and Autopsy in Archaeological Human Skeletal Remains. *Journal of Archaeological Science: Reports* 3:73–79.

Dupras, Tosha L., and John J. Schultz

2014 Taphonomic Bone Staining and Color Changes in Forensic Contexts. In *Manual of Forensic Taphonomy,* edited by James Pokines and Steven Symes, pp. 315–340. CRC Press, Boca Raton, Florida.

Efremov, Ivan A.

1940 Taphonomy: A New Branch of Paleontology. *Pan-American Geologist* 74(2):81–93.

Fernández-Jalvo, Yolanda, Peter Andres, Dolores Pesquero, Colin Smith, Dolores Marín-Monfort, Begoña Sánchez, Eva-Maria Geigl, and Antonio Alonso
2010 Early Bone Diagenesis in Temperate Environments Part I: Surface Features and Histology. *Palaeogeography, Palaeoclimatology, Palaeoecology* 288:62–81.
Harrington, Judith M., and Robert L. Blakely
1995 Bones in the Basement: Bioarchaeology of Historic Remains in Nonmortuary Contexts. In *Bodies of Evidence: Reconstructing History through Skeletal Analysis,* edited by Anne L. Grauer, pp 105–120. Wiley-Liss, New York.
Hodge, Christina J.
2012 Non-Bodies of Knowledge: Anatomized Remains from the Holden Chapel Collection, Harvard. *Journal of Social Archaeology* 13(1):122–149.
Hodge, Christina J., Jane Lyden Rousseau, and Michèle E. Morgan
2017 Teachings of the Dead: The Archaeology of Anatomized Remains from Holden Chapel, Harvard University. In *The Bioarchaeology of Dissection and Autopsy in the United States,* edited by Kenneth C. Nystrom, pp. 115–142. Springer International, Heidelberg, Germany.
Kendall, Christopher, Anne Eriksen, Ioannis Kontopoulos, Matthew Collins, and Gordon Turner-Walker
2018 Diagenesis of Archaeological Bone and Tooth. *Palaeogeography, Palaeoclimatology, Palaeoecology* 491:21–37.
López-Costas, Olalla, with Óscar Lantes-Suárez, and Antonio Cortizas
2016 Chemical Compositional Changes in Archaeological Human Bones due to Diagenesis: Type of Bone vs Soil Environment. *Journal of Archaeological Science* 67:43–51.
Lusignan, Kimberly A.
2004 A Historical and Osteological Analysis of Postmortem Medical Practices from the Albany County Almshouse Cemetery Skeletal Sample. Master's thesis, Department of Anthropology, State University of New York at Albany.
McFarlin, Shannon C., Lawrence F. Wineski
1997 The Cutting Edge: Experimental Anatomy and The Reconstruction of Nineteenth-Century Dissection Techniques. In *Bones in the Basement: Postmortem Racism in Nineteenth-Century Medical Training,* edited by Robert L. Blakely and Judith R. Harrington, pp.107–161. Smithsonian Institution Press: Washington, DC.
Novak, Shannon, and Wesley Willoughby
2010 Resurrectionists' Excursions: Evidence of Postmortem Dissection from the Spring Street Presbyterian Church. *Northeast Historical Archaeology* 39(1):134–152.
Nystrom, Kenneth C.
2011 Postmortem Examinations and the Embodiment of Inequality in the 19th Century United States. *International Journal of Paleopathology* 1(3–4):164–172.
2014 The Bioarchaeology of Structural Violence and Dissection in the 19th-Century United States. *American Anthropologist* 116(4):765–779.
Owsley, Douglas W
1995 Contribution of Bioarchaeological Research to Knowledge of Nineteenth-Cen-

tury Surgery. In *Grave Reflections: Portraying the Past Through Cemetery Studies,* edited by S. R. Saunders and A. Herring, pp. 121–151. Canadian Scholars' Press, Toronto.

Pokines, James, and Joan E. Baker

2013 Effects of Burial Environment on Osseous Remains. In *Manual of Forensic Taphonomy,* edited by James Pokines and Steven A. Symes, pp. 73–114. CRC Press, Boca Raton, Florida.

Solano, Martin C.

2006 The Life Stresses of Poverty: Skeletal and Historical Indicators of Activity Patterns in the Albany County Almshouse Skeletal Collection, 1825–1925. PhD dissertation, Department of Anthropology, State University of New York at Albany.

Symes, Steve A., Erin N. Chapman, Christopher W. Rainwater, Luis L. Cabo, and Susan M. T. Myster

2010 *Knife and Saw Toolmark Analysis in Bone: A Manual Designed for the Examination of Criminal Mutilation and Dismemberment.* Mercyhurst College, Erie, Pennsylvania.

9

Analysis of Commingled Human Remains and Element Representation from the Point San Jose Collection

MARIA COX, VALERIE SGHEIZA, SAMUEL A. MIJAL, KRISTEN A. BROEHL, HEATHER MACINNES, AND ERIC J. BARTELINK

The analysis of the number of individuals and element representation in potential medical waste deposits provides a window into both the context of the assemblage and the individuals included within. The Point San Jose assemblage demonstrates many challenges common to medical waste features. Remains were thoroughly commingled, highly fragmented, and recovery rates of individual elements were variable. These factors can obscure information about the feature and hinder analysis.

These types of assemblages are often highly complex, therefore, the research design necessitated multiple avenues through which to approach assessment of commingling. To this end, our team established three main goals for our analysis. The first was to ascertain the number of individuals represented in the Point San Jose burial feature. To add to these data, we sought to characterize fragmentation patterns and element representation in order to contribute to interpretations of depositional context. Our final goal was to assess which combination of commingling analysis methods were the most appropriate for this collection and would result in usable data.

As demonstrated in the analysis of remains from the Medical College of Virginia (Owsley and Bruwelheide 2012), a detailed inventory of element counts, both complete and fragmentary, was necessary before beginning any further analysis. The applicability of commingled methodology varies based upon the degree of commingling, fragmentation, and recovery. With this challenge in mind, calculation of the Minimum Number of Individuals (MNI) was an essential starting point (White 1953). When the condition of the remains was adequate for pair-matching, the Most Likely Number of Individuals (MLNI) was also assessed in order to estimate the original number of individuals present in the assemblage (Adams and Konigsberg 2004).

The results of this initial analysis permitted the calculation of element representation and proportions. These analyses were conducted to assess patterns which could provide information on selection and disposal of body parts during the creation of this assemblage. An analysis of skeletal element representation can provide important data on proportions of individuals represented even when the original number of individuals in the assemblage is unclear (Owsley and Bruwelheide 2012). The degree of fragmentation and commingling may also serve to contextualize the treatment of remains prior to and during disposal (Hodge et al. 2017).

In addition to these approaches, portable X-ray fluorescence (pXRF) spectrometry has shown promise in segregating individuals based on chemical elemental concentrations in skeletal elements, especially for small-scale commingling situations. Portable XRF is a nondestructive method of analyzing trace elements in a material and has proven useful in a number of applications. Most recently it has been applied to human commingling cases in forensic and archaeological contexts (Finlayson et al. 2017; Gonzalez-Rodriguez and Fowler 2013; Perrone et al. 2014).

The Point San Jose assemblage is hypothesized to be a result of anatomical dissection disposal. We tested this hypothesis by comparing element recovery rates and proportions with those of a collection with similar context: the Holden Chapel Collection. Holden Chapel is a medical waste deposit associated with Harvard Medical School dating from 1800–1850 (Hodge et al. 2017). Comparing Point San Jose to a commingled burial with the same probable context allowed for comparisons of patterns between assemblages, which could provide additional information about the Point San Jose feature's context.

Methods

Approaches to Commingling

MNI is the most common method used to establish the minimum number of individuals present in an assemblage. Establishing the MNI involves sorting remains by element and side. In a traditional MNI, the highest count of an element by side is the MNI (White 1953). Although a traditional MNI relies on information from one element at a time, age- and sex-informed MNI considers the sex and age variation within a collection (Bokonyi 1970; Grayson 1984). Dividing an assemblage into sex and age categories results in greater overall MNI values because element frequencies are unlikely to be the same across all age and sex categories. Unlike spatial categories, such as an excavation unit, age and sex categories avoid the problem of double counting individuals because one individual cannot belong to multiple categories (Grayson 1984). MNI is also a direct function of recovery rate. As recovery rate decreases, MNI decreases proportionately, making it a biased estimator of the original number of individuals (Adams and Konigsberg 2008). Using age estimations to establish MNI reduces bias by more accurately estimating the number of individuals in the assemblage by creating smaller categories for comparison. See Sgheiza and Willey (this volume) for details on sex and age estimation methodology.

To better estimate the original number of individuals represented, the most likely number of individuals (MLNI) may be calculated instead of MNI (Adams and Konigsberg 2004). This method relies on pair-matching and focuses on symmetry between antimeres in size, shape, and morphological features, such as joint surfaces and muscle attachment sites (Adams and Konigsberg 2004, 2008). The number of matched pairs are compared to the number of rights and lefts of that element. This model is based on the premise that paired elements belong to the same individual, while unpaired elements each belong to separate individuals (Seber 1982). The MLNI mathematical formula is based on a hypergeometric distribution. The hypergeometric probability function takes into account the probability of pairing the correct elements together and does not assume a normal distribution (Chapman 1951). Since the MLNI minimum estimate is the same as that element's MNI value, the distribution is often skewed to the left, making highest density regions more accurate than confidence intervals (Adams and Konigsberg 2004). The estimate includes slightly over 95 percent of the distribution area in order to preserve the 0.05 alpha level. The highest density range is based on a discrete distribution, meaning that the range omits decimals because portions of a person cannot exist, which results in an estimate slightly exceeding 95 percent.

MLNI calculation attempts to correct the downward bias of MNI. MLNI is a more accurate indicator than MNI of the original number of individuals represented in an assemblage. That being said, MLNI underestimates the original number of individuals from collections with recovery rates below 30 percent (Adams and Konigsberg 2004, 2008). In addition, MLNI requires pair-matching. With high fragmentation or poor preservation, pair-matching may be difficult and can lead to unreliable estimates. Generally, if it is not possible to compute MLNI in an accurate and efficient manner, MNI is used instead of MLNI because its problems are well understood, and it does not overestimate absolute number of individuals (Adams and Konigsberg 2004). As the Point San Jose series could represent disproportionate disposal practices, greater MLNI values must be considered. For this collection, elements with the least fragmentation were selected so that both MNI and MLNI could be calculated and compared.

Osteometric sorting quantitatively evaluates differences in measurements between skeletal elements in commingled assemblages based on the premise that elements from a single individual are likely to be similar in size, shape, and robusticity. This method identifies when the metric variation between elements exceeds the range of normal human variation, thus reducing the number of possible matches within an assemblage (Byrd and Adams 2003). There are multiple statistical models employed in osteometric sorting. These models include those used to compare homologous elements as well as multivariate analyses that compare differences among multiple elements (Byrd 2008; Byrd and LeGarde 2014; Thomas et al. 2013). Osteometrics provide an efficient approach for sorting large commingled collections where there are a great number of potential matches (Byrd and Adams 2003). A full description of the osteometric methods applied to this collection is available in MacInnes (2017) and Broehl (2018).

A thorough analysis of both qualitative and quantitative commingling methods was undertaken by MacInnes (2017), using the Point San Jose collection, to establish the most effective methods, combination of methods, and order of application for collections that share similar attributes. Although visual pair-matching produced the lowest percent of possible matches, it also provided the greatest number of confident matches, indicating this approach as the most effective single method of sorting. When quantitative methods were applied before qualitative, there was a high percentage of exclusion and a low percentage of possible matches to test visually. Using quantitative methods before qualitative methods is, therefore, the most effective sequence of applications to assure a high percentage of confident matches.

Recovery Rates

Recovery rates are calculated as the number of fragments recovered of an element divided by the expected number of that element. The expected number of an element is based on what would be expected anatomically to be present given the MNI of the collection. All reported recovery rates are maxima because the number of expected elements is based on the MNI, which has downward bias. In short, because MNI is directly related to recovery rates, it often underestimates the original number of individuals present. Recovery rates are maxima because they are based on a potential underestimate and so have fewer expected elements in the denominator of the proportion. Since MNI is likely to underestimate the original number of individuals present in this collection, MLNI is also used to calculate expected values to illustrate the range of recovery rates possible per element. Expected counts are modified based on the age variation within the collection. For example, the Point San Jose collection includes fetal remains that would have been too young to have developed some skeletal elements. Element proportions are calculated by dividing the number of fragments of an element by the total number of fragments within the collection. Vertebral and rib fragment counts were excluded when calculating the overall recovery rate due to their high degree of fragmentation.

In addition, the recovery rates presented here have major limitations. Individuals in the Point San Jose assemblage are commingled and elements are highly fragmented, so it is necessary to include all fragments of an element in the recovery rate calculation. Their inclusion means that fragments of a single element may be counted more than once. This step artificially inflates the recovery rates, especially in elements more prone to fragmentation (Grayson 1984). Despite this complication, calculating recovery rates permits cross-comparisons between collections as well as providing information concerning the proportion of an element within the assemblage itself.

Statistical Analyses

Statistical analyses were used to compare element representation in Point San Jose with the Holden Chapel assemblage, as well as between areas of the body. For these analyses, a z-test of proportions was used to compare proportions of element fragments to the total number of fragments in the collection. Comparing actual fragment counts against expected values demonstrates the interplay between fragmentation and element attrition.

High fragment counts may indicate low attrition, but very high counts consisting mostly of small, fragile fragments, is more indicative of extensive fragmentation. Testing fragment counts also keeps the units similar across each site. Both sites experienced a similar degree of fragmentation. Similar fragmentation levels allow for some control over differences due to fragmentation and preservation. Chi-square tests of association and Cramer's V effect size tests were used to assess element association with preservation. All analyses and visualizations were performed in R 3.5.0 (R Core Team 2015) using the *SuppDists* package (*v1.1–9.4;* Wheeler 2016) in the production of the highest density region (HDR) plots.

X-Ray Fluorescence Spectrometry

Human bone remodels continuously during an individual's lifetime, remodeling completely every 10–20 years (Kini and Nandeesh 2012). As bone remodels, it incorporates trace amounts of elements into its inorganic matrix. Elemental concentrations within bone are influenced by many activities, including in vivo consumption of food and water, metabolic functions, chemical exposure and even inhalation (Perrone et al. 2014). Interindividual variation in trace element concentrations suggests that two bones with very different element concentrations likely originate from different individuals. However, homogeneous trace element concentrations between bones is not necessarily indicative of origin from the same individual. Diagenetic processes *may* homogenize trace element concentrations as bone and depositional environment approach equilibrium (Burton and Price 2000), although this has yet to be rigorously tested. For this reason, XRF techniques should be considered an exclusionary method for analyzing commingling remains. In this analysis, if chemical elemental concentrations between pair-matched skeletal elements were similar between antimeres, the pXRF results provided additional support for their association.

X-ray fluorescence spectrometry operates by bombarding the surface of an element with X-rays. When this bombardment occurs, many X-rays transmit or scatter through the scanned material, but some are absorbed by atoms (Perrone et al. 2014). When X-rays are absorbed, they cause atoms to become ionized by striking and displacing electrons (Pollard et al. 2007). If an X-ray strikes the inner-shell electrons of an atom with enough force to dislodge them, vacancies are created, causing the atom to become unstable. To restore stability, outer-shell electrons fill the inner-shell positions. The movement of electrons from the outer-shell to the inner-shell releases a lower-energy secondary fluorescent X-ray. These secondary fluorescent X-rays are measured by XRF spectrometry. Measurements can be displayed either as a spectrum overlay, which permits the visual comparison of two or more samples, or as an elemental concentration chart (Perrone et al. 2014).

For this study 344 human skeletal elements from the Point San Jose assemblage were scanned, focusing on possible pair-matched bones (n=18). Elements selected for scanning included all six long bones (humerus, radius, ulna, femur, tibia, and fibula), and the clavicle, scapula, os coxae, frontal, occipital, and zygomatic. For articulated elements, such as the bones of the skull, only a single bone was scanned. Scanning locations were selected based on flatness and cortical bone thickness. Flat areas were preferred to limit the amount of airspace between the X-ray emitter and the bone. Likewise, thick cortical bone was preferred to limit the penetration of X-rays into the airspace of the medullary cavity of long bones. Long and short bones were scanned at the midshaft wherever possible. For irregular and flat bones, scans were taken at points of greatest bone thickness and flatness, which varied based on element. Care was taken to avoid scanning near identification number labels and areas that had been reconstructed with adhesive.

Scanning was conducted using a Bruker Tracer IV Series pXRF machine with S1 PXRF software v. 3.8.30, with an attached Bruker vacuum pump. The portable vacuum unit removes air particles between the pXRF and the sample material that can influence element concentration measurements (Christensen et al. 2012). No filter was utilized for these scans, with the expectation that data from a wider range of chemical elements would be more informative than finer resolution of data for fewer elements. A protective plastic screen was placed between the X-ray emitter and the sample material to protect the equipment and ensure that none of the sample material was transferred to the emitter. Each sample was scanned once, for 180 seconds.

Results

Analysis of the Point San Jose collection included the application of extensive commingling methodology, the results of which are reported here. The methods used, from the more established MNI and MLNI to the less traditional approaches like pXRF and element/proportion representation, allowed the results to be as multi-dimensional as possible. These approaches enabled the incorporation of various viewpoints, which aided in the quest to establish the original purpose of the Point San Jose feature.

Commingled Remains

The traditional MNI of the Point San Jose collection is 22 individuals, based on right tali. The next most frequent MNI is the right first metatarsal (MT1) (n = 19), followed by the right metatarsal 3 (MT3), left radius, left metatarsal 5 (MT5), and right metacarpal 2 (MC2) (all n = 18). Five other elements have MNI counts of 17, and three have counts of 16 (see Table B.2, and other supplementary tables, in the UFDC: https://ufdc.ufl.edu/AA00090501/00013). As there was an unusually high proportion of hand and foot bones in this collection, further analysis was undertaken (Broehl 2018). The inclusion of metatarsals will not be presented here because the severe taphonomic alteration and fragmentation made the metacarpal results more reliable. After the application of the derived osteometric equation created by Broehl (2018), the MNI counts did increase, although not above that of the talus (n=22). Metacarpal 1 (MC1) had an MNI of 14 (7 pairs), MC2 had an MNI of 20 (11 pairs), MC3 had an MNI of 17 (6 pairs), MC4 had an MNI of 11 (4 pairs), and MC5 had an MNI of 12 (2 pairs).

When refining MNI with age and sex, and using multiple elements, the MNI increases to 25 individuals (Table 9.1). Of these 25 individuals, at least 16 are adults, at least 2 are juveniles, at least 3 are fetal, and 4 individuals are of indeterminate age and sex. Juvenile is defined here as below 18 years of age. Among the adults, eight individuals are estimated to be male based on metric assessment and morphological traits of the tali, os coxae, and humeri. Six others are estimated to be female based on humerus metrics. Two additional adults have ambiguous sex estimations based on the os coxae. Of the juveniles, at least one individual is aged 7–15 years and at least one is 15–18 years. These age intervals are estimated using epiphyseal development and fusion. Three fetal left petrous portions are present. As no fetal tali are present, and there are at least 16 adults and 2 juveniles present, the number of tali in the collection indicates that there are 4 additional people present. Unfortunately, no sex or age estimations can be inferred from development or metric analysis of these four individuals.

The MLNI values range from 19 (humerus) to 38 (os coxae), with highest density regions showing an overall range of 18 to 69 individuals. MNI estimation for this assemblage indicates that at least 25 individuals are present, so values above 25 are emphasized. Table 9.2 indicates the element MNI, potential pairs using osteometric sorting, pairs established with visual matching, and the final MLNI with associated HDR. Further osteometric analyses can be found in MacInnes (2017) and Broehl (2018). The MLNI

Table 9.1. Point San Jose age- and sex-informed minimum number of individuals (MNI) (Element MNIs in parentheses)

Age	Sex	Element	MNI
Adult	Male	Talus (8), Os coxae (8), Humerus (8), Femur (6)	8
Adult	Female	Humerus (6), Os coxae (4), Femur (2)	6
Adult	Ambiguous	Os coxae (2)	2
15–18	Indeterminate	Radius, Metatarsal[a]	1
7–15	Indeterminate	Calcaneus, Radius[a]	1
Fetal	Indeterminate	Temporal, Frontal, Tibia	3
Indeterminate	Indeterminate	Talus	4
		Total	25

[a] Several juvenile elements present in the assemblage could belong to two or more age intervals. They are omitted from this table. Included here are the elements that demonstrate mutually exclusive age intervals.

equation permits inclusion of multiple elements. Estimates based on multiple elements have greater precision because they increase the degrees of freedom, which decreases the variance in the equation. The humerus, tibia, and os coxae were chosen based on anatomical separation, meaning that recovery is likely independent. Previous research has shown that elements articulating at a joint have a greater tendency to be recovered together contrasted with those with at least one intervening element (Sgheiza 2018). Using these three elements, the MLNI value is 30, with a likelihood region of 25 to 40 (see Figure A.1, HDR= 95.3%, and other supplementary figures, in the UFDC: https://ufdc.ufl.edu/AA00090501/00001). As multiple elements produce a more reliable result, the interval of 25 to 40 is used as the final MLNI estimate.

Table 9.2. Point San Jose minimum number of individuals (MNI), matched pairs, and most likely number of individuals (MLNI) with highest density regions (HDR) by elements

Element	MNI	Potential Pairs[a]	Pairs	MLNI (HDR)
Humerus	L=15 R=12	38	9	19 (18–28)
Ulna	L=17 R=13	25	11	20 (19–25)
Radius	L=18 R=13	25	9	25 (22–38)
Femur	L=14 R=10	14	6	22 (18–42)
Tibia	L=8 R=16	14	3	37 (26–69)
Os coxae	L=16 R=13	40	5	38 (28–69)
Humerus + Tibia + Os coxae				30 (25–40)

[a] MacInnes 2017.

Element Recovery and Proportions

Point San Jose recovery rates vary greatly by element (Figure 9.1, Table B.3 in the UFDC: https://ufdc.ufl.edu/AA00090501/00014). Recovery rates for long bones are all high and vary from 74 to 98 percent. The highest recovery rate among the long bones is the radius (98%) and the lowest is the femur (74%). The skull has a recovery rate of 49 percent, distantly followed by teeth (21%). The scapula, vertebrae, ribs, and sternum recovery rates exceed 100 percent. This overrepresentation likely results from greater fragmentation of these elements. Phalanges have a low recovery rate (20%), as well as carpals (26%). On the other hand, metacarpals (52%), metatarsals (60%), and tarsals (54%) all have recovery rates around 50 percent. The overall recovery rate for Point San Jose is 38 percent when using the MNI estimate, and 32 percent when using the MLNI estimate.

Proportions of element fragments from different sides and anatomical regions were also assessed (Figure 9.2A). There was no difference between the number of fragments from the left and right sides. Comparison of the axial and appendicular skeletal elements was inconclusive. Although there were over twice as many fragments from the axial skeleton, there was also a significant relationship with preservation. Axial fragments were significantly more likely to have poor preservation scores, with a Cramer's V score of 0.75 indicating a strong relationship. This result may indicate that there were more identifications because these elements had fragmented into more pieces, rather than from initial abundance. As with side, there was little difference between upper and lower limb element abundance.

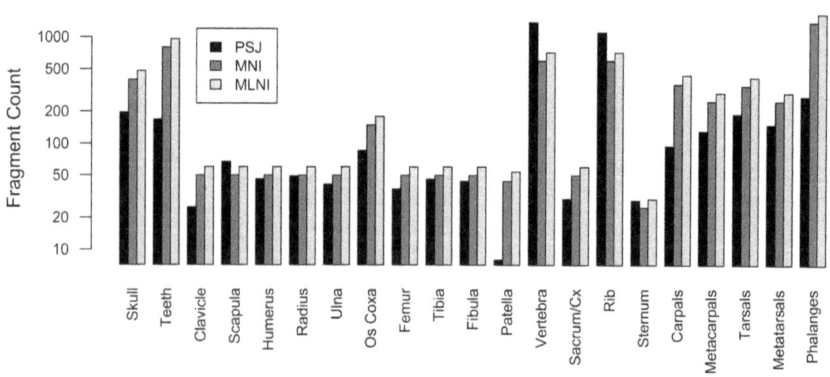

Figure 9.1. Point San Jose (PSJ) fragments observed (log scale), compared with expected number based on MNI and MLNI.

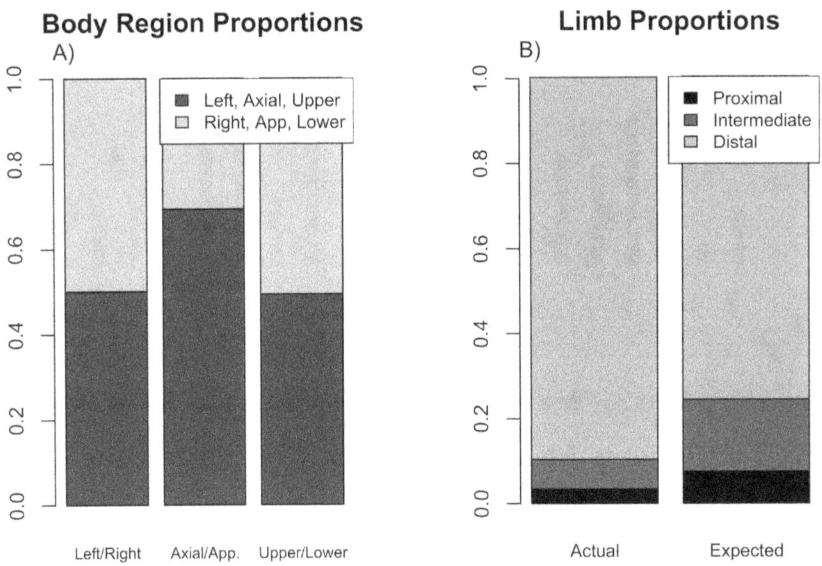

Figure 9.2. Point San Jose fragment proportions by (*A*) body region and (*B*) limb position.

Location along the limb was also analyzed (Figure 9.2B); hands and feet were characterized as distal; radius, ulna, tibia, and fibula as intermediate; and humerus and femur as proximal. Fragments in these categories were divided based on the expected element total. Distal fragments were 16.8 percent less abundant than expected. Intermediate fragments were 10.1 percent more abundant and proximal fragments were 4.1 percent more abundant than expected. These results may be a product of fragmentation because smaller elements may prove less identifiable after fragmentation.

The proportions of element fragments were compared between the Point San Jose and Holden Chapel assemblages (Figure 9.3). The proportion of skull fragments in the Point San Jose collection was not statistically different from the Holden Chapel collection (Table B.4 in the UFDC: https://ufdc.ufl.edu/AA00090501/00015). The teeth, scapulae, humeri, tibiae, radius, and sternum fragments were also not statistically different between collections. Point San Jose and Holden Chapel, however, had significantly different proportions of ulnae, os coxae, femora, fibulae, vertebrae, ribs, clavicle, patella, and all hand and foot bones. Recovery rates were compared across sites, but comparison of fragment proportions was more useful because recovery rates are based on expected values.

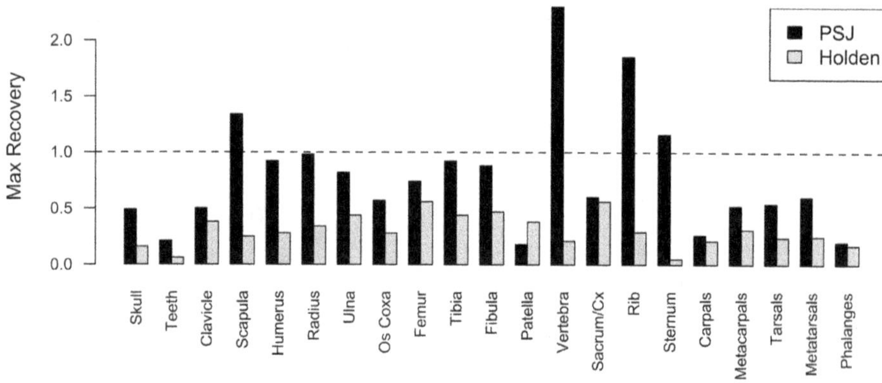

Figure 9.3. Point San Jose (PSJ) recovery rates (based on expected values) compared with Holden Chapel.

X-Ray Fluorescence Spectrometry

Comparisons of calcium and phosphorus concentrations were conducted using a series of bivariate plots for the following elements: femur (Figure A.2), tibia (Figure A.3), fibula (Figure A.4), humerus (Figure A.5), ulna (Figure A.6), radius (Figure A.7), os coxae (Figure A.8), scapula (Figure A.9), and clavicle (Figure A.10). Although the pXRF detected 47 different chemical elements, the current analysis focuses on calcium and phosphorus because these two elements have been shown to be useful for sorting commingled remains (Perrone et al. 2014). Overall, these plots demonstrate that pair-matched elements show no association in the chemical concentration values for either calcium or phosphorus. The plots of the femur (n=5), tibia (n=4), fibula (n=2), humerus (n=3), ulna (n=9), os coxae (n=7), and clavicle (n=3) pair-matches associated no (0%) left-right pairs more closely to each other than nonpaired elements. The plot of radius pair-matches (n=6) associated 1 (16.67%) left-right pair more closely to each other than nonpaired radii. Only a single set of pair-matched scapulae were available for this analysis, precluding a more rigorous evaluation. However, the high variation in calcium and phosphorus values between this pair suggest that a larger sample size would result in their exclusion as pair-matches. Therefore, the data do not provide support for the pair-matched elements.

Discussion

The Point San Jose age- and sex-informed MNI is 25. It consists of 16 adults, as well as 3 fetuses and 2 adolescents. Of the adults, there were eight males, six females, and six individuals of indeterminate sex. The MLNI values vary from 19 to 38 with highest density regions of 18 to 69, but values above the MNI of 25 are emphasized. We were able to select three elements with a lesser degree of fragmentation and with anatomical separation, making that estimate the most reliable. The MLNI is then established at 30, with a highest density region of 25 to 40. The maximum likelihood upper bound of 40 individuals cannot be discounted because this collection may include disarticulated portions of many individuals instead of complete skeletons.

The recovery rates for the Point San Jose collection follow some interesting trends. Large and dense bones tend to preserve better through time and are, therefore, more likely to be recovered (Galloway et al. 1997). This tendency holds true with Point San Jose elements, with long bones recovery rates between 74 percent and 98 percent. The femur, however, holds the bottom of that range (74%). The skull has an even lower recovery rate (49%). Although there are many small and more fragile bones in the skull, portions of the cranium and mandible are very dense, which usually results in higher recovery rates. Although teeth have a low recovery rate (21%), their absence may result from antemortem tooth loss, rather than postmortem processes (see Milligan et al., this volume). The sternum, scapulae, vertebrae, and ribs have recovery rates greater than 100 percent. These high frequencies likely result from fragmentation rather than preferential inclusion in the burial feature. This assertion is supported by the significantly poorer preservation of these elements. The high recovery rates of hand and foot bones is also of note. Analyses of metacarpals indicate that there are a high number of pairs and unmatched metacarpals that articulate with a pair. This indicates that there are more individuals represented by two hands than by one in this collection (Broehl 2018).

The lower recovery rates of elements like the femur and skull may provide support for preferential selection of certain elements before disposal. Since Dr. Edwin Bentley, a surgeon attached to the Point San Jose hospital, was an anatomist and collector (Bentley 1870; see Gavette, this volume), skulls and femora may have been selectively removed and curated for use as anatomical or pathological specimens. Cut and saw marks on the bones indicate that the remains were modified during autopsy or dissection (see Peters et al., this volume), further supporting this hypothesis. To test the hypothesis that this feature results from autopsy and preferential selection before disposal, the Point San Jose element composition was compared with another commingled collection with a function similar to that hypothesized for Point San Jose. Holden Chapel and Point San Jose have similar proportions of skull, teeth, humeri, and tibiae fragments. In Holden Chapel, the low representation of the skull is attributed to retention for use as anatomical or pathological specimens (Hodge et al. 2017). This explanation may also be applicable to the Point San Jose materials.

Point San Jose element proportions were significantly different from Holden Chapel in several categories, including the femur, ulnae, scapulae, sternum, vertebrae, and ribs. The differences in vertebrae, rib, scapulae, and sternum proportions are likely a product of the high fragmentation in these elements in the Point San Jose collection rather than a true representation of difference. Likewise, the difference in proportions in the ulnae between Point San Jose and Holden Chapel likely results from fragmentation. This conclusion was reached because the proportion of radii fragments are similar between Point San Jose and Holden Chapel, and the probability of these bones being recovered together is high (Sgheiza 2018). The significant difference in the proportion of femur fragments is of particular note. Although the recovery rate for the femur is high (74%), it is lower than expected. Bentley was known for his skill with amputations, particularly at the hip. There was evidence of probable practice amputations within the Point San Jose collection, which may provide an explanation for missing femora.

It is important to note that element representation patterns may be obscured or altered by excavation practices. Owsley and Bruwelheide (2012) note that recovery rates of remains from the Medical College of Virginia track closely with element size, indicating that smaller elements were likely lost in backdirt due to their lower visibility. The femur and elements of the skull, however, are large, dense, and robust to taphonomic change. These factors suggest disposal decisions rather than excavation as the explanation for their reduced recovery relative to other bones with those characteristics.

The most likely explanation for the pXRF data is that the chemical concentrations in the Point San Jose skeletal elements have been affected by diagenesis, probably from the long period of burial, which allowed for ionic exchange between the skeletal remains and surrounding soil and groundwater. The variation between elements in this collection indicates that chemical alteration, likely via diagenesis, has obfuscated the relationship between paired elements. Diagenesis is a stochastic process that can both cause interindividual differences in chemical concentrations to homogenize and intraindividual concentrations to diverge. This process is undoubtedly a confounding factor and one that requires further analysis. Future research should explore more complex statistical methods that use data from multiple chemical elements to build a more distinctive chemical signature of each skeletal element compared.

Although it is possible that the data represent mostly false positive pair-matches, the robusticity of the commingled analysis, as well as the number of false positives this would represent, makes this conclusion unlikely. A third possibility could be that the variation in chemical concentration between correctly pair-matched elements is greater than previously realized, and therefore, pXRF methods cannot be applied reliably. While the application of pXRF to commingled assemblages is relatively new, it has been successfully applied in several contexts (Finlayson et al. 2017; Perrone et al. 2014; Gonzalez-Rodriguez and Fowler 2013). Previous applications do suggest, however, that pXRF methods are most effective when discriminating between a small number of individuals with high interindividual skeletal element representation. The high MNI/MLNI and low intraindividual element representation among the remains from the pit likely contributed to the lack of meaningful association within pairs.

Conclusion

At least 25 individuals are represented in the Point San Jose assemblage. The most likely number of individuals is 30, with an interval of 25 to 40. As this interment includes portions of individuals, rather than complete skeletons, and is extensively commingled, the maximum likelihood of 40 individuals cannot be discounted. The overall recovery rate is low, but anomalies among individual element recovery rates, such as the femur and skull, may indicate preferential selection of elements before disposal. Diagenesis precluded reassociation of elements via pXRF. Further research is needed to clarify whether there are specific chemical elements or combinations of chemical elements that are robust to diagenetic effects under the burial conditions seen here.

Comparisons between the Point San Jose and Holden Chapel assemblages revealed that these two collections are similar in key categories of element proportions. Holden Chapel is a known medical waste deposit associated with dissection; this evidence, combined with the presence of cut and saw marks on the material, supports the conclusion that the Point San Jose feature was utilized for a similar purpose. Analysis of element representation plays a vital role in contextualizing features containing fragmented and commingled remains. These data contributed to other aspects of analysis such as demography, as well as offered more immediate insights through comparative study of features with known histories. Although the route is less direct than the analysis of intact burials, the knowledge gained nevertheless provides meaningful information on complex collections that often include marginalized individuals.

Acknowledgments

The authors would like to acknowledge Peter Gavette, P. Willey, and the National Park Service for their efforts in coordinating this project. Valuable edits to multiple drafts of this manuscript were provided by P. Willey and Colleen Milligan. We also express our gratitude to the student workers who contributed their time and effort to collect the data that is presented in this chapter and for making the full extent of this project possible.

References Cited

Adams, Bradley J., and Lyle W. Konigsberg
2004 Estimation of the Most Likely Number of Individuals from Commingled Human Skeletal Remains. *American Journal of Physical Anthropology* 125(2):138–151.

Adams, Bradley J., and Lyle W. Konigsberg
2008 How Many People? Determining the Number of Individuals Represented by Commingled Human Remains. In *Recovery, Analysis, and Identification of Commingled Human Remains,* edited by Bradley J. Adams and John E. Byrd, pp. 241–255. Humana Press, Totowa, New Jersey.

Bentley, Edwin
1870 A Remarkable Specimen of Puerperal Osteophytes. *Pacific Medical and Surgical Journal* 4:1–5.

Bokonyi, Sandor
1970 A New Method for the Determination of the Number of Individuals in Animal Bone Material. *American Journal of Archaeology* 74(3):291–292.

Broehl, Kristen A.
2018 *Osteometric Sorting of Metacarpals and Metatarsals in Commingled Human Skeletal Assemblages.* Master's thesis, Department of Anthropology, California State University, Chico.

Burton, J. H., and T. D. Price
2000 The Use and Abuse of Trace Elements for Paleodietary Research. In *Biogeochemical Approaches to Paleodietary Analysis,* edited by S. H. Ambrose and M. A. Katzenberg, pp. 159–171. Kluwer Academic/Plenum Publishers, New York.

Byrd, John E.
2008 Models and Methods for Osteometric Sorting. In *Recovery, Analysis, and Identification of Commingled Human Remain,* edited by Bradley J. Adams and John E. Byrd, pp. 199 220. Humana Press, Totowa, New Jersey.

Byrd, John E., and Bradley J. Adams

2003 Osteometric Sorting of Commingled Human Remains. *Journal of Forensic Sciences* 48:717–724.

Byrd, John E., and Carrie B. LeGarde

2014 Osteometric Sorting. In *Commingled Human Remains: Methods in Recovery, Analysis, and Identification,* edited by Bradley J. Adams and John E. Byrd, pp. 167–192. Academic Press, New York.

Chapman, Douglas George

1951 *Some Properties of the Hypergeometric Distribution with Applications to Zoological Sample Censuses* Vol. 1. University of California Publications in Statistics, University of California Press, Berkeley.

Christensen, A. M., M. A. Smith, and R. M. Thomas

2012 Validation of X-ray Fluorescence Spectrometry for Determining Osseous or Dental Origin of Unknown Material. *Journal of Forensic Science* 57:47–51.

Finlayson, J. E., E. J. Bartelink, A. Perrone, and K. Dalton

2017 Multimethod Resolution of a Small-scale Case of Commingling. *Journal of Forensic Sciences* 62(2):493–497.

Galloway A., P. Willey, and L. Snyder

1997 Human Bone Mineral Densities and Survival of Bone Elements: A Contemporary Sample. In *Forensic Taphonomy: The Postmortem Fate of Human Remains,* edited by W. D. Haglund and M. Sorg, pp. 295–317. CDC Press, Boca Raton, Florida.

Gonzalez-Rodriguez, J., and G. Fowler

2013 A Study on the Discrimination of Human Skeletons Using X-ray Fluorescence and Chemometric Tools in Chemical Anthropology. *Forensic Science International* 231(1–3): 407e1.

Grayson, Donald K.

1984 *Quantitative Zooarchaeology: Topics in the Analysis of Archaeological Faunas.* Academic Press, Orlando, Florida.

Hodge, C. J., J. L. Rousseau, and M. E. Morgan

2017 Teachings of the Dead: The Archaeology of Anatomized Remains from Holden Chapel, Harvard University. In *The Bioarchaeology of Dissection and Autopsy in the United States,* edited by K. C. Nystrom, pp.115–142. Springer, New York.

Kini, U., and N. B. Nandeesh

2012 Physiology of Bone Formation, Remodeling, and Metabolism In *Radionuclide and Hybrid Bone Imaging,* edited by I. Fogelman, G. Gnanasegaran, and H. van der Wall, pp. 29–57. Springer, Berlin.

MacInnes, H.

2017 *Specimens of Morbid Anatomy: A Study of Context and Sorting Methods of Commingled Human Remains from the Point San José Ossuary.* Master's thesis, Department of Anthropology, California State University, Chico.

Owsley, D. W., and K. Bruwelheide

2012 *Artifacts and Commingled Skeletal Remains from a Well on the Medical College of Virginia Campus: Human Skeletal Remains from Archaeological Site 44HE814.*

Virginia Commonwealth University, Office of the President Documents, Richmond, Virginia.

Perrone, A., J. E. Finlayson, E. J. Bartelink, and K. D. Dalton

2014 Application of Portable X-ray Fluorescence (XRF) for Sorting Commingled Human Remains In *Commingled Human Remains: Methods in Recovery, Analysis, and Identification,* edited by B. J. Adams and J. E. Byrd, pp. 145–165. Academic Press, Oxford, United Kingdom.

Pollard, M., C. Batt, B. Stern, and S.M.M. Young

2007 *Analytical Chemistry in Archaeology.* Cambridge University Press, Cambridge, United Kingdom.

R Core Team

2015 *R: A Language and Environment for Statistical Computing.* R Foundation for Statistical Computing, Vienna, Austria. http://www.R-project.org.

Seber, George A. F.

1982 *The Estimation of Animal Abundance and Related Parameters. 2nd edition.* MacMillan Publishing Company, New York.

Sgheiza, Valerie

2018 *Two Factors Affecting Multi-Element Quantification of Commingled Human Skeletal Assemblages.* Master's thesis, Department of Anthropology, California State University, Chico.

Thomas, R. M., D. H. Ubelaker, J. E. and Byrd

2013 Tables for the Metric Evaluation of Pair-matching of Human Skeletal Elements. *Journal of Forensic Sciences, 58*(4):952–956.

Wheeler, Bob

2016 *Supplementary Distributions.* R package version 1.1–9.4. https://cran.rproject.org/web/packages/SuppDists/index.html.

White, Theodore E.

1953 A Method of Calculating the Dietary Percentage of Various Food Animals Utilized by Aboriginal Peoples. *American Antiquity* 18(4):396–398.

10

Age, Sex, and Ancestry Estimations of the People from Point San Jose

VALERIE SGHEIZA AND P. WILLEY

Demography, the statistical study of population parameters, provides essential information about the mortality, growth, and structure of historical populations. Although commingled skeletal assemblages present additional challenges, characterizing age, sex, and ancestry distributions for these individuals is no less important than it is in more typical cemeteries. However, demographic analysis of commingled remains needs to be approached using a revised protocol. Because adult age and ancestry estimations typically rely on different skeletal elements that cannot be readily reassociated in a commingled context, commingling between individuals may necessitate approaching these two demographic variables separately. The number of skeletal elements that may contribute to an age estimation in a child requires further protocol revisions in age-diverse commingled assemblages such as Point San Jose. Despite these differences, we maintain that a thorough demographic analysis of the Point San Jose assemblage is both achievable and informative for understanding the assemblage.

Here we present a descriptive analysis of the age, sex, and ancestry of the Point San Jose assemblage. We then use comparative data from known populations to test the null hypothesis that the Point San Jose assemblage is reflective of the mortality demographics of nineteenth-century California against several alternatives, including military deaths, hospital deaths, catastrophic mortality from the local population, among others. While it is unlikely that we can capture the exact circumstances of the Point San Jose assemblage in our comparative samples, what we do not find is often as interesting and informative as what we do. Our comparative analysis shows evidence of selection of individuals by age but not sex, and possible selection by ancestry. In this analysis we provide a demographic perspective to the questions of who these people were and how they came to be interred together.

Methods

Age Estimation Methods

Subadult age estimations used skeletal and dental growth and development (Scheuer et al. 2000). Duplicate subadult elements and those bones with non-overlapping age estimations indicated different individuals. Fetal age ranges were estimated from duplicate petrous temporal portions using images and descriptions by Scheuer and colleagues (2000). Additional age ranges for older subadults were estimated by consolidating elements into the smallest number of individuals possible with the smallest age ranges consistent with the observed elements. That is, when the developmental states observed were inconsistent with normal development of a single individual, an additional individual was indicated. Once this number was finalized, age ranges were estimated as the smallest intervals required to encompass the observed developmental states.

Adult age estimations relied on pelvic assessments following *Standards for Data Collection from Human Skeletal Remains* (Buikstra and Ubelaker 1994) as well as *Transition Analysis* (ADBOU version 2; Boldsen et al. 2002). Adult age distributions were based on the highest count pubic symphysis or auricular surface from the same side. The auricular surface of the left ilium constituted this greatest number. The age distributions resulting from the two methods were compared; we tested the adult age midpoint based on *Standards* and maximum likelihood ages from *Transition Analysis* using Wilcoxon signed-rank test (paired, nonparametric), and Kolmogorov-Smirnov test (unpaired, nonparametric) in R (R Core Team 2015). We omitted subadult age comparisons from these *Standards-Transition Analysis* comparisons.

Life table statistics employ the age ranges from *Standards,* including juvenile ages. We focused on survivorship (l_x the proportion of the original population that survives to each successive year of age) because it is possible to compare survivorship between groups. It was necessary to compute life table statistics first before proceeding to hazard modeling because our comparative census data are in a life table form. Three fetuses from Point San Jose constitute additional challenges. While the fetal remains merit discussion, they were excluded from comparative testing because most mortality data commence with live births. Rather than compute life table statistics from range midpoints as is frequently calculated, the entire estimated age intervals were used.

Due to the small sample size, an integral approach was used instead of life table categories to derive an age distribution in Point San Jose. Because the probability distribution in each age range was unknown, each year of the age range was assigned an equal probability. Each age interval was then treated as a rectangle with a length extending the distance of the interval, a total area of one, indicating the age range. The age interval rectangles were stacked on each other and integrated year-by-year by summing up the areas by each year represented in the sample. Because the integral for each year represents a number or fraction of individuals, life table equations can be calculated from the resulting integrals. Life table statistics were calculated using equations provided by Chamberlain (2006). Survivorship was plotted in R (R Core Team 2015).

Sex Estimation Methods

Sex estimation of the Point San Jose skeletal material was based on morphological methods that emphasized the innominate (Buikstra and Ubelaker 1994). To compare age distributions between the sexes, sex was established by counting the greatest number of age-diagnostic elements of the same type (ilium or pubis) and side (left or right) for each sex. This count, however, was not from the same element used for the demography by age alone. The age intervals for each sex were then calculated using the year-by-year integration procedures described previously. The resulting distributions were plotted in R (R Core Team 2015) and the sex ratios were compared with other groups using a χ^2 test with Yates continuity correction for small samples.

Except the talus, FORDISC (Jantz and Ousley 2005) discriminant functions estimate sex for some other postcranial elements. For the talus, sex was estimated using Steele's length and width discriminant function (Steele 1976, table 4, function 2). Values above the male mean were identified as males, and values between the male mean and sectioning point identified as probable males. Values below those points could represent either females or juveniles.

The overall sex ratio was established using the greatest number of each sex estimated from the same element and same side. For males, this number was based on the talus count and for females, the humerus provided this estimation.

Ancestry Estimation Methods and Comparisons

Ancestry was the final demographic parameter explored. Because *Standards* lacks ancestral affinity methods, we used FORDISC 3.0 with the Forensic Databank (FDB) comparative groups. In keeping with recommended practices, the ancestry of subadults was not estimated.

Once the Point San Jose skeletal age, sex, and ancestry profiles were estimated, we compared those results with other groups. Comparative groups were chosen to reflect plausible mortality patterns of the individuals in the Point San Jose sample. We compared the age distribution from Point San Jose with the civilian age distribution of California deaths in 1870 (US Census 1872a; see Table B.5, and other supplementary tables, in the UFDC: https://ufdc.ufl.edu/AA00090501/00016) as well as the US military regimental death records of frontier regulars in California, 1860–1875. Billy Markland (personal communication) provided those military data (Markland 2001). We also included nineteenth-century hospital mortality data (Chamberlain 2006). Note that the hospital sample showed selection by age. No deaths under 12 years occurred in that series. The original dataset indicated evidence of age-heaping around 5- and 10-year intervals, which we compensated for by applying a 3-year moving average. This correction method reported the number of individuals at each age as the average number of individuals reported for that year and the two adjacent years. Over-reporting at 5- and 10-year intervals was smoothed systematically while still counting each individual exactly once.

Finally, the 1870 California living census was used as a catastrophic sample; catastrophic mortality populations imitate the demographic structure of the corresponding living population. Survivorship curves were compared graphically. Point San Jose and the hospital sample were further compared using the delta method for estimating standard error of survivorship curves (Seber 1982). Point San Jose was modeled using a Gompertz distribution, and the hospital sample fits a Gompertz distribution for adult mortality and a negative Gompertz distribution for childhood mortality (Gage 1988, Siler 1979). It was important to first find a model that had a good fit to the data before estimating the standard error. This step was required because the delta method estimates standard error from the model parameters, rather than directly from the data itself. If the incorrect model was chosen, then the standard error would reflect the poor fit of the model rather than a realistic estimation of the uncertainty of the survivorship curve.

Point San Jose adult ages estimated with *Transition Analysis* were used for these comparisons. *Transition Analysis* provides larger age ranges than estimations using *Standards*. Aside from the hospital sample, where a three-year moving average was applied, survivorships for the comparative datasets were calculated from the age categories of that specific sample.

Using a χ^2 test for proportions, the Point San Jose sex distribution was compared with the California mortality sex ratio from the 1870 Census (US Census Bureau 1872a). Due to the small number of skulls in the Point San Jose assemblage that were sufficiently complete for ancestry estimation using FORDISC, we ran a power analysis for the χ^2 test but did not run the test itself. The Point San Jose ancestry assessments are discussed in conjunction with California mortality by social race category according to the 1870 census (US Census Bureau 1872b). That census information is summarized in Tables B.6, B.7, and B.8 in the UFDC: https://ufdc.ufl.edu/AA00090501/00017, https://ufdc.ufl.edu/AA00090501/00018, https://ufdc.ufl.edu/AA00090501/00019.

Results

Age Results

The Point San Jose age distribution was summarized using life table statistics (Table B.9). When compared to other samples (Figure 10.1), Point San Jose survivorship was atypical. There were no individuals between birth and seven years, despite the fact that the first five years of life are typically a period of high mortality. While Point San Jose was most similar to the nineteenth-century hospital sample, there were significant deviations in survivorship in young and old adults (Figure 10.1). No significant difference occurred between the adult age distributions based on *Standards* and *Transition Analysis* (Table B.10). This conclusion held true for both Wilcoxon signed-rank (V = 37, p = 0.1167) and Kolmogorov-Smirnov (D = 0.375, p = 0.2106) comparisons.

Sex Results

Four males and four probable males were represented by right ilia. A total of eight males was corroborated by the talus and humerus counts. Based on discriminant functions of the humerus, at least six adult females were represented.

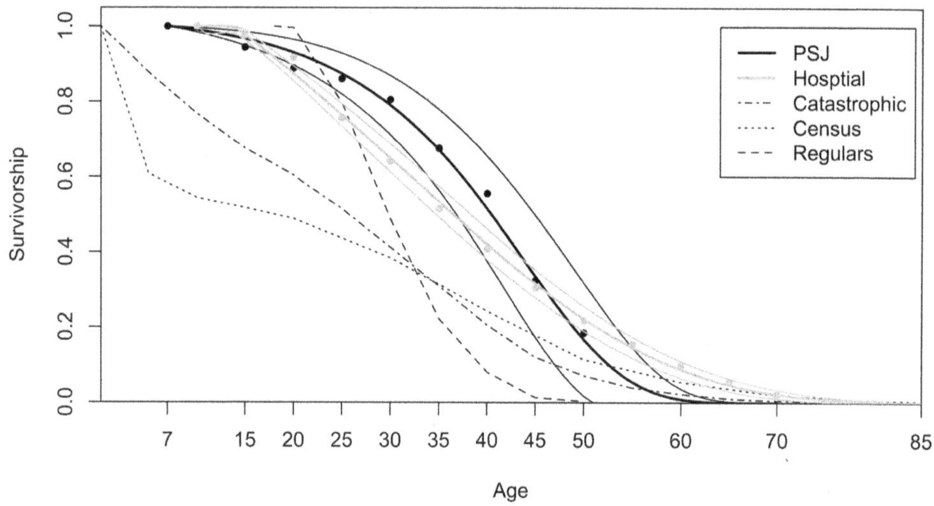

Figure 10.1. Comparison of Point San Jose (PSJ) survivorship (lx) with a nineteenth-century hospital population (Hospital), 1870 California living census (Catastrophic), 1870 mortality census (Census), and nineteenth-century California US military mortality (Regulars). Age in years. Boundary lines for PSJ and Hospital data represent standard error provided by the delta method.

Ancestry Results

Seven Point San Jose crania were complete enough to attempt to estimate ancestry. Posterior probabilities and typicalities suggested ancestries from several groups found within and outside the geographic boundaries of the continental US. Due to the challenges of classifying continuous traits into groups of limited reference sample sizes, artificial geographic boundaries, and no allowances for individuals of mixed ancestry, most individuals had significant posterior probabilities with multiple groups. For individuals with a sex estimate from the cranium, FDB ancestries were Hispanic, American Indian, Japanese, and White. When sex was treated as unknown, additional FDB ancestries also included Black, Vietnamese, and Guatemalan (Table 10.1).

Examining mortality in the California 1870 census by group (US Census 1872b; Tables S 10.2 and S 10.3), "Whites" constituted the vast majority of deaths (93.4%), distantly followed by "Chinese" (4.6%), "Indians" (1.3%), and "Coloreds" (0.7%). In the 1870 census, "Whites" included Europeans, Euro-Americans, Whites, and Hispanics, such as Californios and migrants from Latin America. "Chinese" consisted of people from that country as

Table 10.1. Ancestry assessments of Point San Jose crania using discriminant functions and morphological traits

Catalog #	Sex[a]	FDB both sexes	FDB one sex
42389	Indet	Japanese M .488/.870, Vietnamese and Hispanic Ms. Both ~.25/.75	
42390	F?	Guatemalan M .566/.334, American Indian M .302/.220	White F .599/.289, Hispanic F .258/.170
42391	M?	Black F .87/.739, White F .075/.181	American Indian M .414/.443, Hispanic M .408/.429
42392	F	White F .632/.714, Japanese F .223/.501	
42393	M?	American Indian F .560/.538, Japanese M .296/.396	Hispanic M .427/.435, Japanese M .338/.390
42684	F?	Japanese F .598/.415, White F .307/.283	
43089	M?	American Indian M .707/.959, American Indian F .125/.507	American Indian M .829/.174, Japanese M .080/.040

Note: Groups with highest and second highest posterior probabilities are given. Numbers provided are posterior and typicality probabilities, separated by a forward slash.
[a] Sex abbreviations: F = Female, F? = Probable female, Indet = Indeterminant, M? = Probable male, M = Male.

well as possibly a few Japanese people. "Indians" referred to Native Americans. "Coloreds" in the census encompassed those of African and mixed African and European descent (Hickman 1997). With these limitations in mind, the census categories may be more accurately considered social race categories than ancestry categories because they followed social conventions such as hypodescent rather than the primary geographic origins of a person's ancestors. For example, at the time, someone of primarily European ancestry would have been categorized as "Colored" rather than "White" if they had any known African ancestry (Hickman 1997).

We did not statistically compare FORDISC estimates with the census because we only had an ancestry sample size of seven individuals. Rather than comparing Point San Jose ancestry proportions to those of the racial categories in the census mortality, we performed a power analysis. If we wished to compare the proportions of minority groups between Point San Jose and the census (93.4% "White"), there would have to be fewer than two Point San Jose individuals who would have been considered "White"

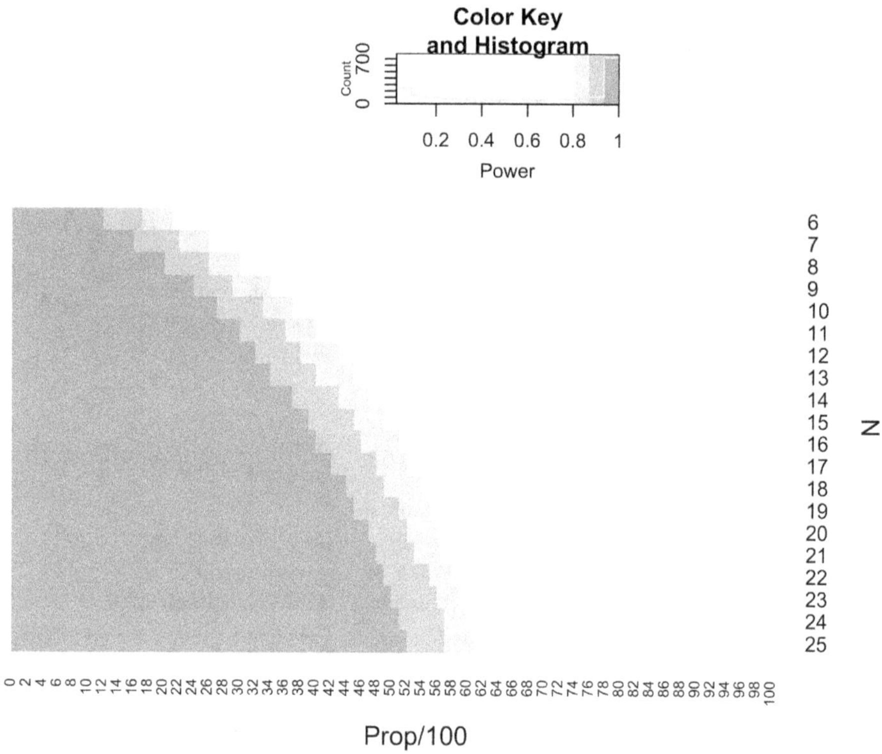

Figure 10.2. Power analysis for chi-square test of proportions using Point San Jose sample size for ancestry. Hypothetical proportion of majority white ancestry in Point San Jose is given on the x-axis and sample sizes on the y-axis. Gray regions indicate proportion + sample size combinations with a power of 0.8 or larger (80% probability of rejecting null hypothesis of no difference in proportions when null hypothesis is false.)

according to the 1870 census (Figure 10.2). Detecting a difference with a White-classifying proportion in Point San Jose of 50 percent would require a sample size of 16 and detecting a difference above 60 percent would require over 25 individuals, larger than the minimum number of individuals for Point San Jose (see Cox et al., this volume). The sample size for Point San Jose ancestry was therefore insufficient to detect meaningful differences in proportions between groups.

Comparisons and Discussion

First, this section compares the Point San Jose materials with other groups by age, sex, and ancestry. Then, it provides a discussion of nineteenth-century cadaver accession and demographic implications for the social context of the Point San Jose skeletal series.

Age Comparisons and Discussion

The Point San Jose age distribution showed significant differences from all the comparative series, especially the military and census mortality, and living census (Figure 10.1). Differences occurred mainly from the lack of individuals between birth and five years in the Point San Jose sample. Large differences occurred between Point San Jose and the California living and California mortality groups. These differences indicated that Point San Jose did not represent a random sample of contemporary Californians. In addition, differences in survivorship, including the presence of children and older adults, and the presence of females demonstrated that Point San Jose did not represent the remains of soldiers alone.

Although still showing a statistically significant difference from Point San Jose, the nineteenth-century hospital sample survivorship was the most similar. One possibility is that, similar to the hospital sample, selection of the Point San Jose assemblage was limited to individuals over a certain age. This possibility might explain the fact that all Point San Jose individuals under five years of age were pre-term fetuses. The Point San Jose fetuses might have been included in the assemblage, carried inside the bodies of pregnant female adults. A second possibility is that the odd age structure of the hospital group resulted from lumping their death ages or our application of the three-year moving average. Using the three-year averages should have made the hospital sample more like Point San Jose, however, because it extends the probability density two years younger to age 10.

Sex Comparisons and Discussion

In the Point San Jose skeletal series, both sexes were represented. This Point San Jose distribution contrasted with military nineteenth-century enlistees, who were all males. On the other hand, the Point San Jose and California mortality census sex distributions did not differ significantly ($\chi^2 = 0.9648$, $p = 0.3260$, Yates continuity correction).

Next, the adult age distribution by sex was tested. The Point San Jose and California mortality census age-sex distributions did not differ significantly ($\chi^2 = 0.9648$, $p = 0.3260$, Yates continuity correction). In contrast, a significant difference in age proportions by sex existed between Point San Jose and the military regulars ($\chi^2 = 108.13$, $p < 0.001$). A caveat should be noted. Because no reliable method exists for estimating adult age from the humerus, only the four adult females represented by the pelvis were considered when comparing age distributions by sex (See Figure A.11, and other supplementary figures, in the UFDC: https://ufdc.ufl.edu/AA00090501/00011), not the other two females represented. They were omitted from this consideration.

Ancestry Comparisons and Discussion

The Point San Jose skeletal series included several possible ancestral groups, most of which were well-represented in nineteenth-century California. It is important to note, however, that ancestry categories in FORDISC are not directly comparable to census classifications for a number of reasons. The 1870 census categories represented socially ascribed and identified races with uncertain biological correspondence. While ancestry is often presented as a less typological and more biologically meaningful alternative to race, this means that ancestry estimates depend on different variables than those used to assign race, so ancestry estimates and census categories provide different information.

It is also important to consider the contextual interpretation of ancestry estimation results, especially when an individual has significant posterior probabilities in multiple groups. FORDISC is a classification tool that uses artificially constructed reference sample categories, many of which correspond to socially mediated, biologically meaningless boundaries (i.e., nation-states). Some of these categories are more racialized than others. For example, "American Black" is a FORDISC category that uses identified forensic cases from the FDB (Ousley and Jantz 1998) as its reference sample. Identification as Black in the US is racially mediated by skin color, which is not directly linked to skeletal traits or proportion of African ancestry (Lee et al. 2010, Relethford 2002).

Regarding the FORDISC software itself, posterior probabilities represent classification probabilities, not ancestry proportions, so having multiple significant posterior probabilities is not indicative of having ancestries from multiple groups. It may mean that the individual has an ancestry not represented in FORDISC, or that their ancestry is represented but the reference sample is too small or their measurements are "atypical." Multiple significant posterior probabilities could also occur if reference samples from two groups are very similar so FORDISC does not distinguish between them well. Even if the individual has mixed ancestry, this heritage will not necessarily correspond to the groups indicated by FORDISC. It is worth mentioning that the concept of "mixed" ancestry is only a question of temporal depth. Human gene flow has been extensive since ancient times. Human variation is continuous, defying discrete categorization; there are no bounded human ancestral groups (Relethford 2009). At best, ancestry estimates in FORDISC assign probabilities of belonging to a particular reference sample relative to other reference samples in the software.

With these challenges in mind, ancestry estimates from Point San Jose still provide valuable information about the persons in the assemblage. While we cannot say that there is evidence of selection by ancestry, the majority of Point San Jose individuals with estimated ancestries had posterior probabilities corresponding to minority groups that would have faced significant disadvantages in historical California (i.e., the colonial abuses toward California Indians). It is certainly possible that ancestry was a factor that influenced the probability of each person ending up in the Point San Jose assemblage, likely mediated by related factors such as socioeconomic status.

Summary

This chapter explored the demographics of the Point San Jose assemblage, including age, sex, and ancestry. Once described, we compared those skeletal assessments with US Army soldiers' mortality profile, the 1870 California living and mortality census, and a nineteenth-century hospital group. Most comparisons showed pronounced differences. Point San Jose survivorship did not match the typical pattern of a reproductively viable population. This difference was likely due to the unusual absence of individuals between birth and five years. The Point San Jose skeletons included the very young (fetuses) and middle-aged adults but lacked juveniles (notably infants and children) that a typical mortality census would have in large numbers. Consequently, the Point San Jose age distribution contrasted with the military, 1870 California mortality, and the 1870 California living census. Only the nineteenth-century hospital group showed some similarities with the Point San Jose age distribution. This similarity in age distribution suggested that both groups excluded normally frequent younger subadults. Similarities in survivorship with the hospital sample suggest a hospital as a possible source for decedents who ultimately ended up at Point San Jose (see Gavette, this volume).

The Point San Jose adult skeletons included both females and males. The Point San Jose sex ratios showed statistical similarities to the 1870 mortality census both by sex and sex-by-adult-age. Results of this comparison suggested that neither sex was preferentially selected for dissection.

We do not have a sufficient sample size to test for statistically significant evidence of selection of individuals by ancestry. That said, most individuals in Point San Jose showed ancestry estimates corresponding with groups that would have faced discrimination and other disadvantages in historical California. Considering Point San Jose ages and ancestries together, they did not represent a random selection or disposal of human remains. Our analysis suggests that younger adults and possibly other lower socioeconomic groups were preferentially exploited for the purposes of dissection and identification of pathology in remains. Our findings are consistent with an individual systematically collecting remains for medical study (see Peters et al., this volume).

References Cited

Boldsen, Jesper L., George R. Milner, Lyle W. Konigsberg, and James W. Wood

2002 Transition Analysis: A New Method for Estimating Age from Skeletons. In *Paleodemography: Age Distributions from Skeletal Samples,* edited by Robert D. Hoppa and James W. Vaupel, pp. 73–106. Cambridge University Press, Cambridge, United Kingdom.

Buikstra, Jane E., and Douglas H. Ubelaker

1994 *Standards for Data Collection from Human Skeletal Remains.* Arkansas Archaeological Report Series, 45. Arkansas Archaeological Survey, Fayetteville.

Chamberlain, Andrew T.

2006 *Demography in Archaeology.* Cambridge University Press, Cambridge, United Kingdom.

Gage, T. B.

1988 Mathematical Hazard Models of Mortality: An Alternative to Model Life Tables. *American Journal of Physical Anthropology* 76(4):429–441.

Hickman, Christine B.

1997 The Devil and the One Drop Rule: Racial Categories, African Americans, and the US Census. *Michigan Law Review* 95(5):1161–1265.

Jantz, Richard L., and Stephen D. Ousley

2005 FORDISC 3: Computerized Forensic Discriminant Functions. Version 3.1.35. math.mercyhurst.edu/~sousley/Fordisc/Help/Fordisc3_Help.pdf, accessed August 9, 2016.

Lee, Yin Leng, Susan Teitelbaum, Mary S. Wolff, James G. Wetmur, and Jia Chen
2010 Comparing Genetic Ancestry and Self-Reported Race/Ethnicity in a Multiethnic Population in New York City. *Journal of Genetics* 89(4):417.
Markland, Billy
2001 Database of US Military Frontier Regular Deaths. http://freepages.history.rootsweb.ancestry.com/~familyinformation/frontierregs.xls, accessed August 2016.
Ousley, Stephen D., and Richard L. Jantz
1998 The Forensic Data Bank: Documenting Skeletal Trends in the United States. In *Forensic Osteology: Advances in the Identification of Human Remains,* edited by Kathy Reichs, pp. 441–458. C. C. Thomas, Springfield, Illinois.
R Core Team
2015 R: A Language and Environment for Statistical Computing. R Foundation for Statistical Computing. Vienna, Austria. http://www.R-project.org.
Relethford, J. H.
2002 Apportionment of Global Human Genetic Diversity Based on Craniometrics and Skin Color. *American Journal of Physical* Anthropology 118(4):393–398.
2009 Race and Global Patterns of Phenotypic Variation. *American Journal of Physical Anthropology* 139(1):16–22.
Scheuer, Louise, Sue Black, and Craig Cunningham
2000 *Developmental Juvenile Osteology.* Academic Press, London.
Seber, G.A.F.
1982 *The Estimation of Animal Abundance and Related Parameters,* 2nd edition. Charles Griffin, London.
Siler, William
1979 A Competing-Risk Model for Animal Mortality. *Ecology* 60(4):750–757.
Steele, D. Gentry
1976 The Estimation of Sex on the Basis of the Talus and Calcaneus. *American Journal of Physical Anthropology* 45(3):581–588.
US Census Bureau
1872a *1870 Census: Volume 1. The Statistics of the Population.* US Department of Commerce, Washington, DC.
1872b *1870 Census: A Compendium.* Government Publishing Office, Washington, DC.

11

Point San Jose Statures as Indications of Stress

P. WILLEY

Stature provides insights into the lives of individuals and populations, serving as a proxy for health and well-being. Adult stature results from the combined, synergistic effects of genetics and environmental conditions that occur during a person's growth. Genetic anomalies aside, some authorities assert that genetics' influences are minimal, so they claim that mean statures in a population mostly reflect environmental influences and adaptability (for a contrary view, see Weedon and Frayling 2008).

If stature is influenced by environment, then height can be used as an indicator of stress and well-being during growth and development (Eveleth and Tanner 1976; Bogin 1999). Diet, malnutrition and undernutrition, metabolism, illness and health, epidemiology, disease loads, standard of living, and socioeconomic status appear to have the most environmental impact on growth. Unless compensatory catchup growth occurs following growth delay (Boersma and Wit 1997), prolonged, poor prenatal and postnatal conditions produce shortened adult statures (Beydoun and Saftlas 2008; Dancause et al. 2012). Advantageous conditions, on the other hand, enhance more growth and lead to greater adult heights.

In living groups, adult stature indicates change through time (secular trend), by geographic location, group affiliation, sexual dimorphism, socioeconomic differences, and other living conditions. Those influences are well-established in living individuals and living groups and populations. (For a contrary perspective, see Seckler 1982.)

Similar processes presumably affected past populations and reflected aspects of health and nutritional status. Using stature estimations, bioarchaeologists study stresses on and environments of past groups as general, nonspecific indicators of well-being. To make those inferences in skeletal series, stature must be estimated first. For these applications, stature is usually estimated using limb bone lengths and regression formulae. Limb bone length-to-stature relationship may be influenced by the person's sex, age, ancestry, time period, and exposure to stresses (Trotter 1970).

In the present instance, Point San Jose stature estimations based on limb bone lengths offered insights into those people's lives and living conditions. If the Point San Jose bones represented a downtrodden, oppressed group, then their statures should be shorter when compared with members of more affluent groups.

Methods and Samples

Estimating Point San Jose statures requires several methodological steps and assumptions. Under most circumstances using limb bone lengths to estimate stature relies on other biological assessments; sex, age, and ancestry should be assessed and considered when estimating stature (Moore and Ross 2013:151). Although sex, age, and ancestry could be estimated for some Point San Jose elements, they cannot be established reliably for the limb bones. The commingled, disassociated, disarticulated archaeological context of the Point San Jose skeletal remains causes demographic uncertainty of sex, age, and ancestry estimations for the limb bones employed in stature estimations. Consequently, a general approach is required, one that does not require those demographic parameters.

To deal with stature estimation problems resulting from differences by sex, age, and ancestry, a general regression formula is employed. FORDISC 3.1 offers stature estimation formulae for several ancestral groups and both sexes (Jantz and Ousley 2005). In addition to those formulae for specific sexes and ancestries, it also provides a general means of estimating stature when the skeleton's sex and ancestry are unknown. To estimate stature of the Point San Jose remains, we employed Fordisc's "Any" function for the nineteenth-century Terry Collection cadavers. The "Any" base sample consists of Blacks and whites, females and males, all combined (Jantz and Ousley 2005:45). That software produces stature estimations in inches, which are used for all comparisons throughout this chapter. All of the Point San Jose limb bone length measurements (humerus, radius, ulna, femur, tibia, and fibula) and both sides (left and right) together comprise the overall, all-inclusive estimations.

Descriptive and hypothesis testing statistics employ the Statistical Package for the Social Sciences (Version 20.0). Samples of living individuals are required for comparisons. Although recorded nineteenth-century statures are relatively common east of the Mississippi River (e.g., Margo and Steckel 1983), West Coast statures prove infrequent and have small sample sizes. With those limitations in mind, the comparative male sample comes from Billy Markland's (2001) regimental death records of frontier regulars. From his overall dataset, consisting of more than 11,000 entries, the 215 soldiers who died in California between 1860 and 1875 are selected for their temporal comparability with the Point San Jose series. Their living statures had been measured at the time of their enlistments. As previous studies have noted (Komlos 1992), historical female stature samples are even fewer and prove more difficult to find than males. The most comparable female series is a small sample (n=92) of nineteenth-century white and Black females imprisoned across the entire West (Carson 2011:159, table 2, Far West column). They had been measured at the time of their incarcerations.

Results

Description of Point San Jose All-Inclusive Statures

Combining stature estimations from all Pont San Jose limb bones and both sides, a total of 67 estimations could be calculated (Figure 11.1). Considering all elements with stature estimations by side, lefts (n=35) outnumber rights (n=32). The small difference in side counts, however, falls short of statistical significance.

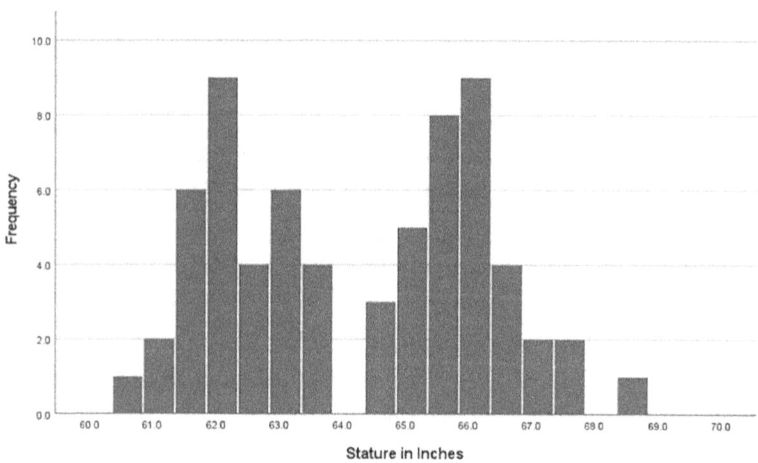

Figure 11.1. All-inclusive Point San Jose stature distribution in one-inch intervals. Note bimodal distribution, with intervening trough (64 inches). Estimations include all limb bones and both sides.

Overall, the all-inclusive stature estimations indicate a group averaging approximately 64.3 inches (in). The estimations vary from a short 60.6 in to a tall 70.1 in. Incidentally, both the short and tall extremes of this range derive from the same side and element (right humerus).

Rather than the expected "normal" or bell-shaped curve, the all-inclusive statures show a striking bimodality distribution (Figure 11.1). One peak occurs at 61–63 in, and the other at 65–66 in with an intervening trough at 64 in. This observation is considered later in the discussion section. The distribution of all-inclusive stature estimations (combining all six elements and both sides) may mask important differences hidden among the overall composite stature distribution. Differences within the series require a separate analysis.

Within-Sample Comparisons of Point San Jose Statures

Here we consider side and element stature estimations separately. They may contain important, nuanced, specific information. When separated by element and side, the stature distributions appear to reveal differences (Figure 11.2, Table 11.1).

The visual impression of the stature distribution, however, belies statistical accuracy. The side of limb bone does not influence estimated stature—neither for any of the limb bones individually nor for the overall distribution (two-tailed p-values from 0.265 to 0.889). A note is in order here.

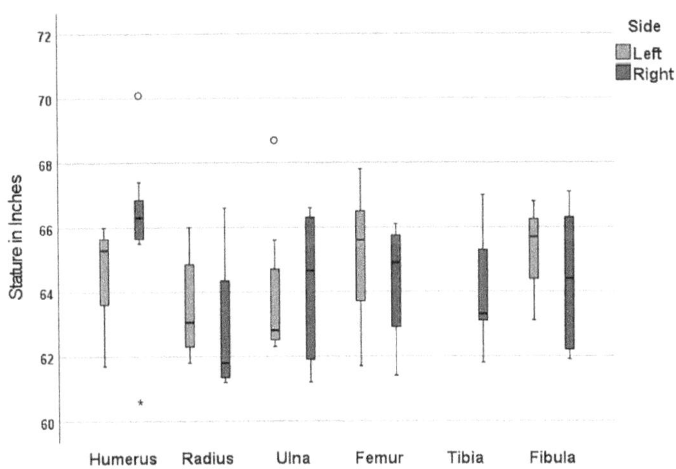

Figure 11.2. Point San Jose statures (in inches) by element and side. Circles represent outliers, asterisk an extreme value. Note: no left tibia lengths were measurable, so by necessity they are omitted from this graph.

Table 11.1. Point San Jose stature means, numbers, and standard deviations (in inches) by element and side

Bone and Side	Mean	Number	Standard Deviation
Fibula, left	65.2000	3	1.90000
Fibular, right	64.3800	5	2.34457
Femur, left	65.1500	6	2.20159
Femur, right	64.3250	4	2.07103
Humerus, left	64.5000	7	1.86279
Humerus, right	66.0000	7	2.84019
Radius, left	63.5250	12	1.50098
Radius, right	62.8500	4	2.52784
Tibia, right	63.9667	6	1.86190
Ulna, left	64.0286	7	2.36623
Ulna, right	64.2167	6	2.38782
Total	64.3313	67	2.14266

Note: No left tibiae lengths could be measured for stature estimation and so are omitted from this table as well as related comparisons.

No complete left tibiae survived for length measurement that the stature estimation requires. Consequently, tibiae had to be omitted from this comparison of sides.

If element side has no effect on stature estimations, perhaps differences occur among the elements themselves. Comparing variation in Point San Jose statures among the six limb bones also fails to achieve significance ($F = 0.953$; $df = 10, 1$; $p = 0.493$). Similarly, the side-element interaction is not significant ($F = 0.662$; $df = 4$; $p = 0.621$).

Summarizing these results, stature estimations by sides and elements do not differ from one another. The apparent differences noted earlier (Figure 11.2) are not attributable to side, element, or side-element interaction.

Comparison of Point San Jose Statures with Other Groups

Point San Jose stature estimations are compared with nineteenth-century California male military and Far West female prisoner statures as well as worldwide living groups. For these comparisons, the Point San Jose element and side with the greatest number is used (left radius, n=12).

The difference between Point San Jose and nineteenth-century military statures is statistically significant (t ~25.76, $df = 226$, $p < 0.000$). Point San Jose stature estimations average about 3.5 in shorter than the military casualties' mean stature. The Point San Jose heights are so short, in fact, that using the 1874 minimum stature enlistment standard (65 in; Ewing 1894:179–180), only one (8.3 percentage) of the Point San Jose stature estimations based on radii meets the minimum enlistment height. The other 91.7 percent would have been disqualified from enlistment based on this stature requirement. Such shortness may reflect differences in ancestry, migration, socioeconomic status, and/or sex.

The difference between Point San Jose and nineteenth-century Far West female prisoner (63.0 + 2.9 in) statures do not appear to be statistically significant. Point San Jose stature estimations average 1.3 in taller than the Far West female prisoner mean stature.

A worldwide comparison provides a broader perspective than the US military and Far West prisoners alone. Continuing to use the left radius stature mean estimation (63.53 in = 161.4 cm), the Point San Jose sample is classified as average height when compared with living worldwide samples (Comas 1960:315). The mean Point San Jose stature estimation stands well above pygmoid group averages, such as the Efe Pygmies (143 cm) and Andamaese (148 cm). On the other hand, the Point San Jose statures average well below taller groups, such as Swedes (174.4 cm) and the Sara people of Chad (181.7 cm). The Point San Jose mean stature rests solidly near the midpoint between those four extreme groups.

Discussion

This section assesses the Point San Jose stature estimations for plausible explanations and discusses limits of the results. Specifically, the bimodal stature distribution and limitations of the approaches are evaluated.

As noted earlier, the all-inclusive Point San Jose stature distribution (combined estimations of all elements and both sides) shows a marked bimodality. That stature dispersal contrasts with the normal, bell-shaped distribution expected in most stature distributions. Based on comparisons within the sample (see previous section), neither side nor element cause the bimodal distribution. What other possible causes could produce the bimodality? Four explanations come to mind.

First, the all-inclusive bimodal distribution may be caused by the vicissitudes associated with a small sample size. Point San Jose's moderate sample size (n = 67) may exclude the spectrum of variation actually present in the population from which the sample is drawn. If the sample size were greater, perhaps additional individuals in the middle of the distribution (ca. 64 in) would have been represented and would have smoothed the double-humped distribution into a single, bell-shaped curve. This explanation, considering the moderate sample size, however, seems unlikely.

Second, bimodality may result from the rounding employed in this chapter (Figure 11.1). By chance, the process may have rounded the all-inclusive, whole-inch (xx.0–xx.9 in) statures into intervals other than those in the trough (ca. 64 in). To test this claim, the stature estimations are re-drawn lumping the distribution into intervals with smaller measurement intervals than the one previously presented in Figure 11.1. When graphed at smaller, half-inch intervals (i.e., xx.0–xx.4 in and xx.5–xx.9 in), the all-inclusive stature estimations still show a deficit in the 64.5–64.9 in interval (n=1), similar to the 1-inch bimodal distribution. When graphed at larger, inch-and-a-half intervals (e.g., 60.0–61.4 in, 61.5–62.9 in, . . . 69.0–70.4 in), the all-inclusive stature estimations show a flattened normal curve in the 63–66-in intervals, although lacking the marked hiatus at 64 in. Based on results of these additional intervals, the all-inclusive depression about 64 in appears to be real, not an artifact of graphing distortion.

Third, the all-inclusive Point San Jose bimodal stature distribution might be explained by different ancestries being represented—a taller ancestral group and a shorter one. Much variation exists in statures by ancestral group. Consider, for instance, the enormous differences between the Efe Pygmies and the Sara stature averages mentioned above; 15.4 in (39 cm) separate their mean statures (Comas 1960). Based on demographic conclusions previously presented (see Sgheiza and Willey, this volume), several ancestral groups are probably represented at Point San Jose. If one of those ancestral groups is taller and the other shorter, they may have produced the bimodal stature distribution observed.

Fourth, the bimodal stature distribution might have resulted from sexual dimorphism in the Point San Jose sample. Greater genetic differences, after all, exist between the sexes than among ancestries. If this explanation is correct, then the Point San Jose stature distribution displays a taller male sample and a shorter female one with little overlap between the two sexes. The hypothetical Point San Jose females average 62.5 in tall and hypothetical males average 66.3 in. These hypothetical mean statures are roughly comparable to the nineteenth-century prisoner females (63.0 in) and military males (67.8 in) employed earlier.

Two approaches to assessing the bimodal possibility exist. One approach to assess the possibility of Point San Jose sexual dimorphism is to use a subset of the all-inclusive stature estimations for those limb bones showing the most diagnostic sexual dimorphism. Unfortunately, the two Point

San Jose limb bones with the greatest sexual dimorphism have samples too small to show a realistic stature estimation distribution: humeri (left n = 7) and femora (left n = 6).

Another approach to assessing possible Point San Jose sexual dimorphism exists. Does the amount of sexual dimorphism in the Point San Jose sample reflect that found in other groups? That question can be addressed using a worldwide sample. Employing 64.5 inches as the dividing point between the two hypothetical sexes (females < 64.5 in, males >64.5 in), we can estimate each hypothetical sex's mean stature, then calculate the sexual dimorphism coefficient following a standard formula (Willey 1990:158). Once calculated, the dimorphism coefficient can be compared with a worldwide sample (Gray and Wolfe 1980:442, table 1, Total Sample).

Calculating height dimorphism from Point San Jose's hypothetical means (presented above), the sexual dimorphism coefficient is 5.82. The worldwide sample's average sexual dimorphism coefficient is larger (7.36) than Point San Jose's. The dimorphism that the all-inclusive Point San Jose stature estimations demonstrate falls well within—although somewhat less than—sexual dimorphism of a large, worldwide sample.

In conclusion, whatever the explanation for the all-inclusive Point San Jose stature bimodality, it indicates a heterogeneous sample, representing at least two groups. In addition to the distinct and separate explanations explored above, the bimodality could result from a combination of those four explanations.

Challenges to this work occur. Many limitations affect the accuracy of Point San Jose stature estimations. As noted earlier, being unable to estimate accurately the sex or ancestry of the limb bones constitutes a major limitation. By necessity, a "one-formula-fits-all" approach is required to estimate Point San Jose stature. That approach, however, ignores differences in limb-bone-lengths-to-stature ratios that may exist either by sex or ancestry. That analytical jump proves a necessary, but shaky leap.

Use of all-inclusive stature estimations to represent the Point San Jose heights constitutes another challenge. All-inclusive stature estimations require combining all elements and both sides together. Issues may arise from that step. First, it has been noted (Shuler et al. 2011) that different bones of an individual yield different stature estimations. That discrepancy does not appear in Point San Jose, at least as well as it can be assessed from this commingled series. The second issue is possible duplication. If each Point San Jose individual is represented by one element from one side only, no duplication occurs. But if individuals are represented by multiple elements or both sides of an element, then those persons' statures may be overrepresented in the distribution and distort the group's actual statures. The commingled context of Point San Jose precludes assessing these possibilities in detail.

Summary

Adult stature provides a means of evaluating the health and well-being of people during their growth period. Poor living conditions and illnesses delay growth and ultimately may produce shorter adult statures; good environmental conditions and good health enhance growth and may yield taller adults. In bioarchaeology, skeletal heights are estimated and then compared with statures of other archaeological samples as well as those of living groups. Those comparisons indicate the relative well-being of the groups.

Based on the approach used here, the Point San Jose skeletal series demonstrates average stature, at least from a worldwide perspective. No substantial stature evidence for deprivation or ill health exist on that level. On the other hand, compared with nineteenth-century US military and Far West prison samples, some members of Point San Jose stood shorter, at least from the comparative male average. Based on stature estimations, the Point San Jose skeletal heights do not represent a homogeneous military series and certainly include females.

The all-inclusive Point San Jose stature estimations display a bimodal distribution. That distribution may result from small sample size, graphing distortions, ancestral differences, or sexual dimorphism. The last explanation is the most intriguing.

Stature constitutes one measure of health and illness, but as a single, general stress indicator, it has limitations. Isotopic studies can provide insights into migration and diet (see Bartelink and Hall as well as Hall et al., this volume). Disease frequency and dental growth interruptions are other means of assessing the well-being of a group, both during their growth and development as stature as well as throughout peoples' lives (see Milligan et al., this volume).

Acknowledgments

Billy Markland forwarded links to his western frontier military regular death database. His database chronicles second half-nineteenth century military deaths of enlisted soldiers and volunteers, most occurring west of the Mississippi River. Among those tens of thousands of deaths, several hundred of those deaths happened in California. Billy kindly provided access and permission to use his data for this project.

References Cited

Beydoun, H., and A. F. Saftlas

2008 Physical and Mental Outcomes of Prenatal Maternal Stress in Human and Animal Studies: A Review of Recent Evidence. *Paediatric and Perinatal Epidemiology* 22:438–466.

Boersma, Bart, and Jan Maarten Wit

1997 Catch-up Growth. *Endocrine Reviews* 18(5):646–661.

Bogin, Barry

1999 *Patterns of Human Growth.* 2nd edition. Cambridge University Press, Cambridge, United Kingdom.

Carson, Scott Alan

2011 Height of Female Americans in the 19th Century and the Antebellum Puzzle. *Economics & Human Biology* 9:157–164.

Comas, Juan

1960 *Manual of Physical Anthropology.* Charles C. Thomas, Springfield, Illinois.

Dancause, K. N., X. J. Cao, F. Veru, S. Xu, H. Long, C. Yu, D. P. Laplante, C. D. Walker, and S. King

2012 Prenatal and Early Postnatal Stress Exposure Influences Long Bone Length in Adult Rat Offspring. *American Journal of Physical Anthropology* 149:307–311.

Eveleth, P. B., and J. M. Tanner

1976 *World-Wide Variation in Human Growth.* Cambridge University Press, Cambridge, United Kingdom.

Ewing, Charles B.

1894 The Selection and Physical Examination of the Recruit. In *Transactions of the Third Annual Meeting of the National Guard of the United States,* pp. 166–195. Buxton and Skinner Stationery, St. Louis.

Gray, J. Patrick, and Linda D. Wolfe

1980 Height and Sexual Dimorphism of Stature among Human Societies. *American Journal of Physical Anthropology* 53:441–456.

Jantz, Richard L., and Stephen D. Ousley

2005 FORDISC 3: Computerized Forensic Discriminant Functions. Electronic document, Version 3.1.35. http:// math.mercyhurst.edu/~sousley/Fordisc/Help/Fordisc3_Help.pdf, accessed August 9, 2016. Fordisc Help File 2005:45.

Komlos, John

1992 Toward an Anthropometric History of African-Americans: The Case of the Free Blacks in Antebellum Maryland. In *Strategic Factors in Nineteenth Century*

American Economic History: A Volume to Honor Robert W. Fogel, edited by Claudia Goldin and Hugh Rockoff, pp. 297–329. University of Chicago Press.

Margo, Robert A., and Richard H. Steckel

1983 Heights of Native-Born Whites during the Antebellum Period. *Journal of Economic History* 43:167–174.

Markland, Billy

2001 Database of US Military Frontier Regular Deaths. Markland and Allied Families Resources. Electronic document, http://freepages.history.rootsweb.ancestry.com/~familyinformation/frontierregs.xls, accessed July 4, 2017.

Moore, M. K., and A. H. Ross

2013 Stature Estimation. In *Research Methods in Human Skeletal Biology,* edited by E. A. DiGangi and M. K. Moore, pp. 151–180. Elsevier, Amsterdam.

Seckler, David

1982 Small but Healthy: A Basis Hypothesis in the Theory, Measurement, and Policy of Malnutrition. In *Newer Concepts in Nutrition and Their Implication for Policy,* edited by Pandurang Vasudeo Sukhatme, pp. 127–137. Maharashtra Association for the Cultivation of Science Research Institute, Cornell University, Ithaca, New York.

Shuler, Kristrina A., Marie Elaine Danforth and Jeffrey Auerbach

2011 Challenges in Approaches to Skeletal Stature Estimation: An Example from Prehistoric Eastern Mississippi and Western Alabama. *Southeastern Archaeology* 30(2): 377–389.

Statistical Package for the Social Sciences (SPSS)

2011 *SPSS Statistics for Windows, Version 20.0.* IBM Corporation, Armonk, New York.

Trotter, Mildred

1970 Estimation of Stature from Intact Long Limb Bones. In *Personal Identification in Mass Disasters,* edited by T. D. Stewart, pp. 71–83. Smithsonian Institution, National Museum of Natural History, Washington, DC.

Weedon, Michael N., and Timothy M. Frayling

2008 Reaching New Heights: Insights into the Genetics of Human Stature. *Trends in Genetics* 24 (12):595–603.

Willey, P.

1990 *Prehistoric Warfare on the Great Plains: Skeletal Analysis of the Crow Creek Massacre Victims.* Garland Publishing, New York.

12

Isotopic Perspectives on Residence Patterns at Point San Jose

ERIC J. BARTELINK AND SARAH A. HALL

In the decades following the 1848 Gold Rush, the city of San Francisco experienced significant growth as the population increased from around 100,000 in the 1870s to over 274,000 in the 1880s, facilitated by an expansion in steamboat travel and the establishment of the transcontinental railroad in 1869 (Mosier 2001). Further, the development of local railroads and a ferryboat system connected many of the Bay Area communities (Mosier 2001; Walker 2001). These transportation networks fundamentally changed the demographics of the city and brought a wave of immigration from other parts of the US and abroad.

By 1870, approximately 49.3 percent of San Francisco's population comprised foreign-born individuals (Merriam 1979). Foreign-born immigrants in California were mainly from China, Ireland, Germany, France, England, Mexico, and Canada (Gregory 2020). Living in a major commercial port city, immigrants to San Francisco were often employed in laboring class jobs, especially in manufacturing and trade (Walker 2001). However, by the mid-1870s San Francisco was severely impacted by the nationwide economic depression, resulting in high unemployment. Chinese immigrants in particular became the target of resentment by white residents, paving the way toward passage of the Chinese Exclusion Act of 1882 (Lee 2002). In general, most foreign-born immigrants arrived in San Francisco poor and often suffered from numerous ailments. For example, in 1880, three-fifths of the almshouse population in California was foreign born, which likely reflected their greater degree of poverty coupled with experiencing higher levels of social isolation compared to other Californians (tenBroek 1957). Additionally, from 1860–1880, US-born individuals immigrated to

California from throughout the country, with the largest numbers from New York, Massachusetts, Ohio, Missouri, Pennsylvania, Maine, Illinois, and Iowa (Gregory 2020).

Populations Vulnerable for Anatomical Dissection

In nineteenth-century contexts like Point San Jose, commingled body parts interred along with medical waste represent the detritus of anatomical dissections carried out by physicians and their medical students (Blakely and Harrington 1997; Hodge 2013; Hodge et al. 2017; Owsley et al. 2017). Historical records indicate that individuals used for dissection were almost always of low socioeconomic status and from disadvantaged ethnic groups (Garment et al. 2007; Nystrom 2017; Sappol 2002). These marginalized communities had little power to prevent physicians from taking unclaimed dead bodies from hospitals, almshouses, and poorhouses, or to prevent the practice of grave robbing from local cemeteries (Sappol 2002).

Anatomists and physicians in San Francisco had ready access to unclaimed bodies for medical dissection from public institutions, negating the need for grave robbing (Gibbons 1867). Initially, only the bodies of state prisoners or executed criminals could be used for dissection. However, amendments to California's Anatomy Act in 1866 included other groups and permitted dissection of unclaimed individuals who died in the custody of a public hospital, county prison, or county poorhouse, in lieu of a burial at the county's expense (Hittell 1870). Further, the revised Anatomy Law of 1870 allowed unclaimed bodies to be taken for dissection after as little as 24 hours after death (Meyer 1930). Given the influx of people into San Francisco during the late nineteenth century, it is likely that immigrants to the region (both foreign and US born) would have been particularly vulnerable to anatomical dissection following death. Recent immigrants to California were overrepresented among the almshouse and poorhouse populations; they tended to have fewer social ties to the local community than local residents (half of all adults in California were single males); and they often lacked the financial resources to cover their own burial expenses (tenBroek 1957).

Sourcing Anatomized Remains from Point San Jose

Here, we use both oxygen and strontium isotopes of tooth enamel to investigate whether anatomized individuals interred at Point San Jose likely represent immigrants born in other areas of the US or abroad versus individuals born in the San Francisco Bay Area. In addition, we also examine oxygen and strontium isotopes of bone bioapatite to identify whether these individuals had recently arrived in the Bay Area or if they were more long-term residents in adulthood. Given the greater disparities in health, social networks, and socioeconomic status that recent immigrants experienced relative to local residents, we expect that they would have been more vulnerable to anatomical dissection following death compared to long-term Bay Area residents.

Oxygen and strontium tooth enamel signatures reflect residence patterns during childhood. Enamel isotope values that are consistent with the San Francisco Bay Area would indicate that an individual was likely born in the local area. However, oxygen and strontium tooth enamel signatures that are inconsistent with the Bay Area would indicate that an individual was most likely born in another US region or abroad. In contrast, oxygen and strontium isotope data from cranial bone reflect more recent adult residence patterns. An individual with bone isotope signatures consistent with the Bay Area likely represents someone who lived locally for approximately a decade or more prior to death based on average remodeling rates for human bone (Manolagas 2000). In contrast, bone isotope signatures that are inconsistent with the Bay Area would likely represent a more recent arrival because their skeletal tissues retained isotope signatures from their previous place of origin. We recognize that isotopic signatures that are consistent with the Bay Area do not preclude the possibility that an individual may have been from another location due to the issue of equifinality with isotope data. Thus, we are assuming that an individual *is likely* local to the area if both oxygen and strontium isotopes are consistent with the Bay Area.

Oxygen and Strontium Isotopes

An individual's skeletal tissues provide an isotopic record of food and water consumption patterns, and different skeletal tissues (e.g., teeth vs. bones) correspond to different periods in an individual's life. For example, a permanent tooth provides isotopic information on diet and residence patterns during childhood when that tooth formed. In contrast, bone isotopes reflect an average of approximately 10–20 years of diet and residential location because bone tissue continuously remodels throughout the lifespan (Hedges et al. 2007; Manolagas 2000; Pate 1994). The bioelement oxygen and the geo-element strontium in particular provide powerful geolocation tools for assessing migration patterns from human skeletal remains.

Stable isotopes are variants of the same chemical element that contain the same number of protons and electrons, but a different number of neutrons (Fry 2006). These elemental variants are considered stable because they do not undergo radioactive decay. The standard convention for reporting stable isotope composition is based on a ratio of ratios, and this convention places all measurements on the same scale. The δ-notation is used to transform small ratio values to larger numbers for ease of comparison, calculated as: $\delta = (R_{sample}-R_{standard})/R_{(standard)} \times 1000$.

Stable oxygen ($\delta^{18}O$) isotopes reflect precipitation patterns, with higher values found along the coast and near large bodies of water, and lower values found in more inland and high-altitude settings (Bowen and Wilkinson 2002; Craig 1961; Dansgaard 1964; Gat 1996). Radiogenic strontium ($^{87}Sr/^{86}Sr$) isotopes reflect the geological age of underlying bedrock, and values are passed up the food chain by plants and the animals that consume them (Bentley 2006; Faure 1998; Faure and Powell 1972). Due to their chemical similarity, strontium can substitute for calcium in hydroxyapatite (bone mineral) in the skeleton. Minimal fractionation occurs in $^{87}Sr/^{86}Sr$ ratios between soils, plants, and animals (Blum et al. 2000; Capo et al. 1998), thus, strontium isotope ratios reflect local bioavailable strontium in animals that consume most of their food locally (Bentley 2006).

For oxygen isotopes of solids, the cretaceous limestone fossil Pee Dee belemnite (PDB) is used as the analytical standard, now referred to as Vienna-PDB (VPDB). For water samples, Standard Mean Ocean Water (VSMOW) is the analytical standard, now referred to as Vienna-SMOW (Coplen 1994; Gonfiantini 1978). Strontium isotope ratios, in contrast, are meaningful to several decimal places and are reported as the raw ratio of ^{87}Sr to ^{86}Sr. Predictive isotope models incorporate spatial reference data derived from water and geological maps to construct isotopic landscapes, known as "isoscapes." Isoscape maps are useful for predicting region of origin for animals, plants, soil, water, and other materials (Bataille and Bowen 2012; Bowen et al. 2007).

Materials and Methods

Because the commingled nature of the assemblage limited the scope of our analysis, only individuals with matching cranial and dental remains were selected. This approach allowed exploration of mobility during the life of a single individual. For the purposes of this study, the "Bay Area" refers to all nine counties that encompass the San Francisco Bay estuary, including Alameda, Contra Costa, Marin, Napa, San Francisco, San Mateo, Santa Clara, Solano, and Sonoma. Although the remains were interred in San Francisco (and presumably these deaths occurred there as well), it is possible that some individuals may have been from a surrounding county near San Francisco. During the late nineteenth century, San Francisco was interconnected with these other Bay Area counties through an elaborate railway and ferry system (Wollenberg 1985). In 1873, more than two million trips were made by ferry between Oakland and San Francisco, which increased to over five million trips by 1877 (Wollenberg 1985).

To assess residence patterns for the eight individuals represented by skull-tooth pairs, isoscape prediction maps for oxygen and strontium were created (Bowen et al. 2007; Bataille and Bowen 2012). Tooth enamel isotope signatures that are inconsistent with the Bay Area would represent individuals who immigrated or moved to the area after their childhood (e.g., born in either another region of the US or abroad). Further, for individuals who recently immigrated to the Bay Area (e.g., within the last few years prior to death), bone isotope ratios would reflect their prior location of adult residence. However, individuals who lived in the Bay Area for at least 10 years (based on average bone remodeling rates) would likely show bone isotope ratios consistent with the local region. These expectations are based on the assumption that these individuals would have primarily consumed locally available food resources (i.e., grown locally) and imbibed locally available water, assumptions that are likely to have been met for late nineteenth-century San Francisco (Hall et al., this volume; Luz and Kolodny 1989).

Bone and tooth samples were selected from partial and intact skulls with direct bone-tooth associations (i.e., tooth still embedded in alveolar bone). Further, maxilla-mandible articulations were evaluated to avoid double-sampling the same individual. From each skull, a bone fragment and either a first or second molar was sampled. Each sample was cleaned with a diamond-studded Dremel bit, then washed first in distilled-deionized water, followed by 95 percent and 100 percent ethanol. For bone bioapatite preparation, samples were mechanically pulverized into a powder using a steel mortar and pestle and sieved through a fine-mesh screen (<200 mm). The collagen fraction was removed through treatment with a hydrogen peroxide solution for 48 hours (Crowley and Wheatley 2014). Diagenetic contaminants were removed using a 1.0 M acetate-buffered acetic acid solution for 24 hours (Koch et al. 1997). Samples were then dried overnight in an oven at 60°C.

All enamel bioapatite samples were submitted to IsoForensics, Salt Lake City, Utah, for sample preparation and analysis. Each tooth was mechanically cleaned to remove surface contaminants and then was ultrasonically cleaned in distilled-deionized water. Approximately 2.0 mg of enamel powder was drilled from each tooth, followed by treatment with a 3 percent hydrogen peroxide solution for 15 minutes. Samples were rinsed in distilled-deionized water, and then treated with 0.1M acetic acid for 15 minutes. Samples were rinsed three times in distilled-deionized water, and then dried overnight in an oven at 60°C. Bone samples were also submitted to IsoForensics for strontium analysis. Bone and tooth samples were digested with concentrated pure nitric acid overnight (95°C for bone and 25°C for enamel). The digested samples were filtered and diluted with ultrapure water prior to measuring strontium concentration. The digested samples were then purified to obtain the strontium fraction.

Stable isotope ratios were measured using isotope ratio mass spectrometry. $\delta^{13}C$ and $\delta^{18}O$ values of carbonate (tooth enamel and bone bioapatite) were measured by continuous-flow isotope ratio mass spectrometry at IsoForensics using a GasBench (Thermo Scientific) interfaced to an isotope ratio mass spectrometer (ThermoFinnigan MAT 253). Strontium isotopes ($^{87}Sr/^{86}Sr$) were measured at the Department of Geology and Geophysics at the University of Utah, Salt Lake City, using a multi-collector inductively coupled plasma mass spectrometer (Thermo Scientific High Resolution Neptune).

During the nineteenth century, all drinking water in the San Francisco area was from local sources, reflecting local precipitation. Thus, isoscape maps for tooth enamel and bone bioapatite were generated from precipitation water reference data from the continental US for the eight skull-tooth pairs (Bowen and Wilkinson 2002). Strontium reference data derive from US geological maps (Bataille and Bowen 2012; Beard and Johnson 2000). For oxygen isotopes of tooth enamel and bone bioapatite, the carbonate $\delta^{18}O$ value was converted from the VPDB scale to the VSMOW scale to follow oxygen isotope values of water using the equation by Daux et al. (2008). The $\delta^{18}O$ value of carbonate was then converted to the phosphate scale for water prediction using the equation by Iacumin et al. (1996). $d^{18}O$ prediction maps added ± 0.5‰ to account for unknown error. Prediction maps for bioavailable $^{87}Sr/^{86}Sr$ were constructed using a lithology-based model (Bataille and Bowen 2012). Human $^{87}Sr/^{86}Sr$ ratios were compared directly to the bioavailable strontium values predicted by the model.

Results

Table 12.1 presents the raw data for oxygen and strontium isotopes for bones and teeth from the eight paired skull-tooth samples. Figure 12.1 plots oxygen and strontium isotope results for the tooth and bone samples. Both tooth enamel oxygen and strontium isotope values are elevated relative to bone bioapatite values (mean enamel $\delta^{18}O = -4.0‰ \pm 1.0$, 1 SD; mean enamel $^{87}Sr/^{86}Sr = 0.70965 \pm 0.00227$, 1 SD; mean bone $\delta^{18}O = -5.9‰ \pm 0.6‰$, 1 SD; mean bone $^{87}Sr/^{86}Sr = 0.70879 \pm 0.00052$). Overall, mean oxygen and strontium isotope values are also more variable for teeth compared to bone, indicating greater diversity in values from childhood compared with values from adulthood (Table 12.1).

Figure 12.2 presents an example of one set of isoscape prediction maps for a tooth and bone sample (No. 43082). The areas shown in blue represent oxygen isotope prediction areas, the areas in red represent strontium isotope prediction areas, and the areas in pink are regions where oxygen and strontium isotope predictions overlap. These areas of overlap represent the most likely regions of origin for this individual. As mentioned previously, individuals are categorized as local if their oxygen *and* strontium isotope

Table 12.1. Oxygen and strontium isotope values for bones and teeth from the Point San Jose assemblage

Sample No. (*Skull*)	$\delta^{18}O_{apat}$ (enamel)[a]	$\delta^{18}O_{apat}$ (bone)	$^{87}Sr/^{86}Sr$ (enamel)	$^{87}Sr/^{86}Sr$ (bone)
42389	−2.1	−6.3	0.71123	0.70943
42392	−4.0	−6.0	0.71116	0.70891
42393	−4.4	−5.6	0.70925	0.70852
42685	−3.3	−5.2	0.70534	0.70791
43082	−4.3	−6.8	0.71113	0.70944
43083	−4.6	−5.5	0.70906	0.70909
43084	−3.5	−5.4	0.70782	0.70853
43089	−5.4	−6.8	0.71220	0.70853
Mean	−4.0	−5.9	0.70965	0.70879
SD	1.0	0.6	0.00227	0.00052

[a] All teeth are first molars except sample no. 42392, which is a second molar.

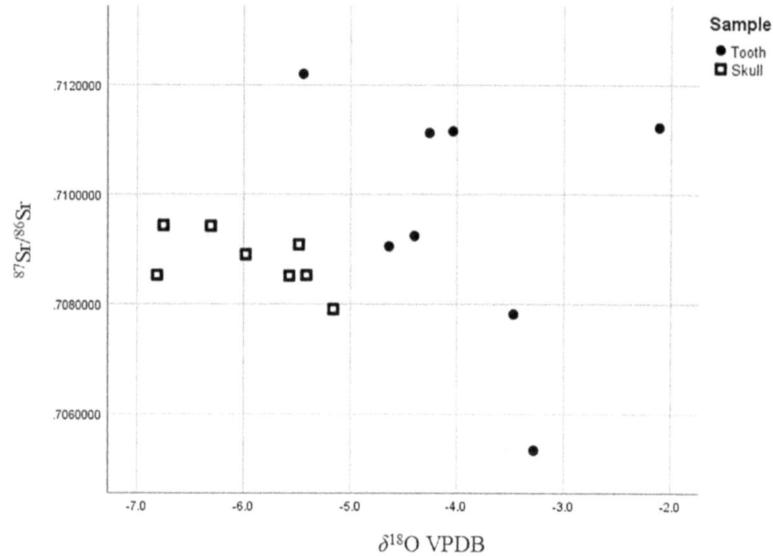

Figure 12.1. Bivariate plot of oxygen and strontium isotope ratios for tooth and bone samples from the skull-tooth pair individuals from Point San Jose.

signatures include one of the Bay Area counties and as nonlocal if either isotope signature falls outside of this region.

For tooth enamel, none of the eight individuals sampled from Point San Jose showed oxygen and strontium isotope values consistent with a childhood residence in the Bay Area (Table 12.2). For seven of these individuals, the nearest prediction areas included southern US states; however, for three of these individuals, isotope values were unable to exclude states from the midwestern US. Only one individual (No. 43089) had possible childhood residence in southern California (consistent with the counties of Santa Barbara, San Luis Obispo, Tulare, Kern, Los Angeles, Riverside, San Bernardino, and San Diego). Because prediction areas were provided for the US only, we cannot be certain whether any of these individuals were born outside of the US.

For bone bioapatite, only one of the eight individuals sampled from Point San Jose showed oxygen and strontium isotope values consistent with adult residence in the Bay Area (No. 43082) (Table 12.2). For this individual, prediction areas included the Bay Area county of Napa, but also the northern California counties of Lake and Tehama and the southern California counties of Kern, San Bernardino, and Inyo. The remaining seven

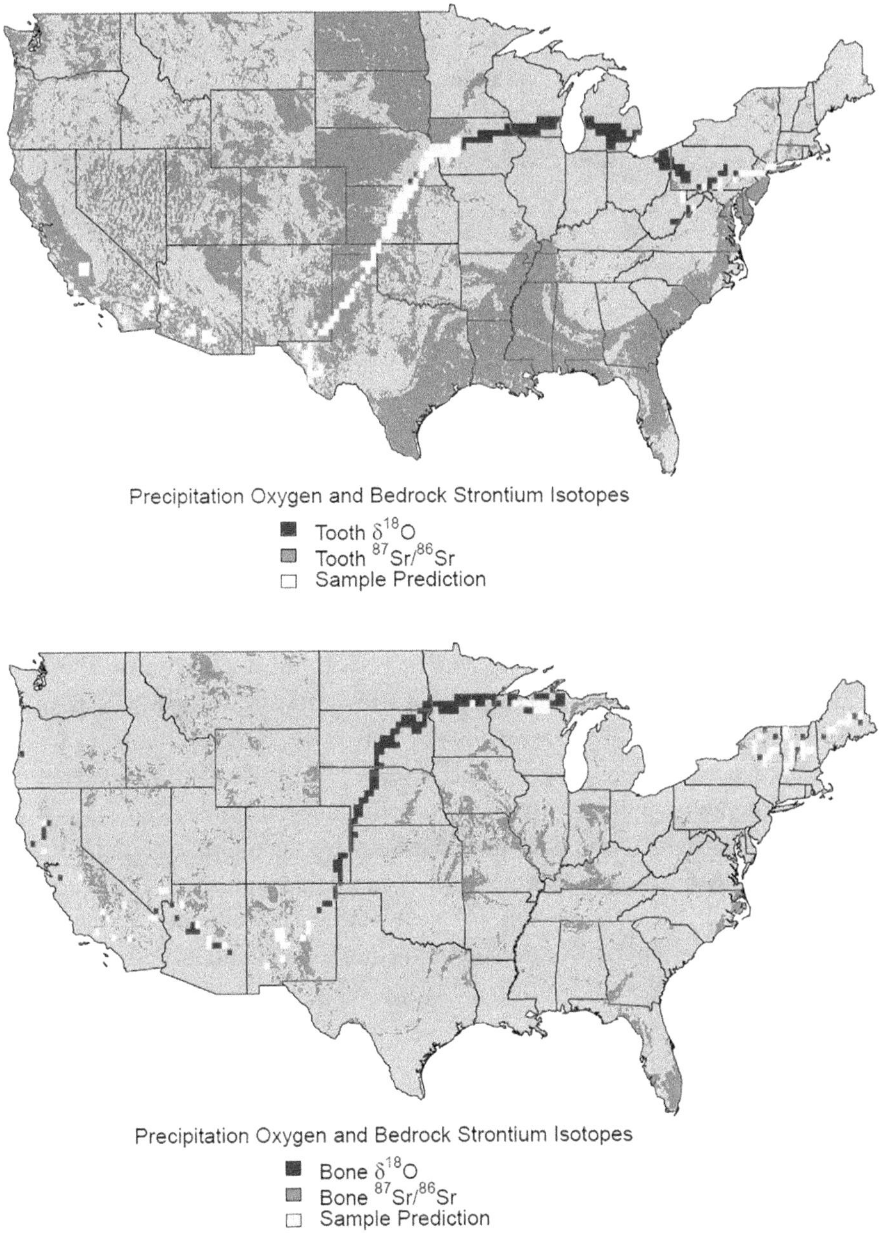

Figure 12.2. Oxygen and strontium isoscape prediction maps for Point San Jose No. 43089 (top map shows tooth enamel; bottom map shows bone bioapatite). The tooth enamel map suggests a possible childhood origin of southern California, although there are many other possibilities within the continental U.S. For bone bioapatite, the prediction map plots the nearest adult residence location in northern California (Colusa, Lake, Tehama, or Yolo County), but also several southern California counties.

Table 12.2. Nearest oxygen and strontium isotope prediction regions based on tooth and bone samples from the Point San Jose assemblage

	Childhood Origin Residence Pattern (Teeth)	Nearest Possible Childhood Region	Adult Origin Residence Pattern Origin (Bone)	Nearest Adulthood Region
42389	Nonlocal	South (Gulf Coast)	Nonlocal	Northern CA
42392	Nonlocal	South	Nonlocal	Central CA
42393	Nonlocal	South	Nonlocal	Southern CA
42685	Nonlocal	South	Nonlocal	Southern CA
43082	Nonlocal	South	Local	Napa County
43083	Nonlocal	South	Nonlocal	Southern CA
43084	Nonlocal	South	Nonlocal	Southern CA
43089	Nonlocal	Southern CA	Nonlocal	Northern CA

individuals had oxygen and strontium isotope values that were inconsistent with residence in the Bay Area in adulthood. However, for all seven of these individuals, isotope values were consistent with other areas in northern and southern California, indicating that while these individuals may have lived outside of the Bay Area, they may still have lived in California (or possibly in other locations) for several years prior to death. Alternatively, these individuals may have lived in more than one location during the past decade of life, which would have averaged isotope signatures acquired from the different residence locations. When compared side-by-side, none of the eight individuals have isotopic signatures that reflect a local Bay Area residence pattern in both childhood and adulthood (Table 12.2).

Discussion

The stable isotope data reported above provide interesting insights on the possible residence patterns of individuals interred in the pit at Point San Jose. It is clear that for all individuals in the sample, oxygen and strontium isotope values reflect possible childhood residence outside of the Bay Area. For seven of these individuals, prediction areas included southern US states, such as Alabama, Arkansas, Florida, Kentucky, Missouri, North and South Carolina, Oklahoma, Tennessee, Texas, and Virginia. With the exception of Missouri, these states did not contribute high levels of immigration to the San Francisco Bay Area (Gregory 2020). However, for three of these individuals, other possible locations include midwestern states such as Illinois and Indiana. Only one individual may have had childhood residence in southern California, although many other regions within the southwest, the south, and the northwest cannot be ruled out.

The bone bioapatite data shed further light on more recent residence patterns for these individuals. The majority of the individuals (seven of eight) were predicted to have lived outside the nine Bay Area counties as adults based on oxygen and strontium results. However, all of these individuals may have had adult residence within either northern or southern California. Only one individual may have lived within the Bay Area as an adult, although other regions within northern and southern California are possible (as well as other parts of the US and abroad).

There are several possible explanations regarding the residence patterns of these eight individuals, none of which are mutually exclusive. The bone isotope data in particular suggest that most of these individuals moved to the San Francisco area from another region of California (or another US region or abroad), perhaps weeks, months, or a few years prior to death. None of these individuals spent their childhood in the Bay Area (based on tooth enamel signatures), but as adults may have lived in other areas of northern or southern California for at least a decade prior to death. This would suggest that these individuals did not live or stay in the Bay Area long enough to incorporate a local signature through long-term consumption of local food and water. If these individuals had recently traveled to San Francisco, they may have had fewer social ties to the local community, and therefore may have been more vulnerable than local residents to dissection.

A second possibility to consider is that individuals may have died outside of the Bay Area, and their bodies were transported to San Francisco to be used in postmortem study. This interpretation would suggest that cadavers for anatomical dissection were in short supply in San Francisco during the late nineteenth century. Given that many restrictions of California's Anatomy Act were removed by the passage of the Anatomy Law of 1870, this seems unlikely (Meyer 1930). However, this possibility cannot be discounted based on the available stable isotope evidence.

A third possibility is that some individuals may have lived in more than one region during the years prior to death, thus having isotopic signatures that were averaged over several different locations, potentially including other countries. This possibility cannot be evaluated using isotopic data and is one of the limitations of using bone isotope data for inferring residence patterns.

Regardless of the precise explanation for the isotopic patterns presented here, this subsample of the individuals at Point San Jose suggests that some of the most marginalized individuals in the region—those who died in local public institutions and were therefore vulnerable to dissection practices—were, in some sense, local, at least in the weeks and month prior to death. All eight sampled individuals demonstrated isotope values that were inconsistent with childhood residence in the Bay Area, and seven of these individuals had isotope values inconsistent with adulthood residence in the Bay Area.

Conclusions

Although the sample size used to infer residence patterns in this study was small (n=8), the results provide important insights regarding the lived experiences of the decedents used for postmortem study at Point San Jose. The oxygen and strontium isotope data suggest that the majority of the individuals examined (seven of eight) did not live in any of the nine Bay Area counties for any extended period prior to death. Thus, these data provide support for the hypothesis that individuals used for anatomical dissection may have engaged in considerable mobility, in both their childhood and adult years. The data are consistent with the hypothesis that some or all of these individuals were foreign born; however, this possibility cannot be tested without evaluating baseline data from other countries. Future studies of the Point San Jose collection should examine isotopic patterns in larger numbers of samples to identify broader patterns of origin among the individuals.

Acknowledgments

The authors thank Lesley Chesson and the staff at IsoForensics, Inc. and UC Davis for their assistance with running samples for mass spectrometry, Evan Kipnis for generating the isotope prediction maps, and the National Park Service's Golden Gate National Park Conservatory for their financial support of this research.

References Cited

Bataille, Clément P., and Gabriel J. Bowen
2012 Mapping $^{87}Sr/^{86}Sr$ Variations in Bedrock and Water for Large Scale Provenance Studies. *Chemical Geology* 304:39–52.

Beard, Brian L., and Clark M. Johnson
2000 Strontium Isotope Composition of Skeletal Material Can Determine the Birth Place and Geographic Mobility of Humans and Animals. *Journal of Forensic Sciences* 45(5):1049–1061.

Bentley, R. A.
2006 Strontium Isotopes from the Earth to the Archaeological Skeleton: A Review. *Journal of Archaeological Method and Theory* 13:135–187.

Blakely, Robert L., and Judith M. Harrington (editors)
1997 *Bones in the Basement: Postmortem Racism in Nineteenth-Century Medical Training.* Smithsonian Institution Press, Washington, DC.

Blum, J. D., E. H. Taliaferro, M. T. Weisse, and R. T. Holmes
2000 Changes in Sr/Ca, Ba/Ca and $^{87}Sr/^{86}Sr$ Ratios between Trophic Levels in Two Forest Ecosystems in the Northeastern U.S.A. *Biogeochemistry* 49:87–101.

Bowen, Gabriel, and Bruce Wilkinson
2002 Spatial Distribution of $\delta^{18}O$ in Meteoric Precipitation. *Geological Society of America Bulletin* 30(4):315–318.

Bowen, Gabriel J., James R. Ehleringer, Lesley A. Chesson, Erik Stange, and Thure E. Cerling
2007 Stable Isotope Ratios of Tap Water in The Contiguous United States. *Water Resources Research* 43(3).

Capo, Rosemary C., Brian W. Stewart, and Oliver A. Chadwick
1998 Strontium Isotopes as Tracers of Ecosystem Processes: Theory and Methods. *Geoderma* 82:197–225.

Coplen, T. B.
1994 Reporting of Stable Carbon, Hydrogen, and Oxygen Isotopic Abundances. *Pure and Applied Chemistry* 66(2):273–276.

Craig, H.
1961 Isotopic Variations in Meteoric Waters. *Science* 133:1702–1703.

Crowley, Brooke E., and Patrick V. Wheatley
2014 To Bleach or Not to Bleach? Comparing Treatment Methods for Isolating Biogenic Carbonate. *Chemical Geology* 381:234–242.

Dansgaard, W.
1964 Stable Isotopes in Precipitation. *Tellus* 16:436–468.

Daux, Valérie, Christophe Lécuyer, Marie-Anne Héran, Romain Amiot, Laurent Simon, François Fourel, François Martineau, Niels Lynnerup, Hervé Reychler, and Gilles Escarguel
2008 Oxygen Isotope Fractionation Between Human Phosphate and Water Revisited. *Journal of Human Evolution* 55(6):1138–1147.

Faure, Gunter
1998 *Principles and Applications of Geochemistry: A Comprehensive Textbook for Geology Students.* Prentice Hall, Upper Saddle River, New Jersey.
Faure, Gunter, and James L. Powell
1972 *Strontium Isotope Geology.* Springer-Verlag, New York.
Fry, Brian
2006 *Stable Isotope Ecology.* Springer, New York.
Garment, Ann, Susan Lederer, Naomi Rogers, and Lisa Boult
2007 Let the Dead Teach the Living: The Rise of Body Bequeathal in 20th-Century America. *Academic Medicine* 82(1):1000–1005.
Gat, Joel R.
1996 Oxygen and Hydrogen Isotopes in the Hydrologic Cycle. *Annual Review of Earth and Planetary Science* 24:225–262.
Gibbons, Henry
1867 An Introductory Lecture to the Fourth Session of Toland Medical College. T. G. Spears & Co. Printers, San Francisco.
Gonfiantini, R.
1978 Standards for Stable Isotope Measurements in Natural Compounds. *Nature* 71(5645):534–536.
Gregory, James
2020 California Migration History 1850–2017. America's Great Migrations Project. https://depts.washington.edu/moving1/California.shtml, accessed May 6, 2020.
Hedges, Robert E. M., John G. Clement, C. David L. Thomas, and Tasmin C. O'Connell.
2007 Collagen Turnover in the Adult Femoral Mid-Shaft: Modeled from Anthropogenic Radiocarbon Tracer Measurements. *American Journal of Physical Anthropology.* 133:808–816.
Hittell, Theodore H.
1870 The General Laws of the State of California, from 1850 to 1864, Inclusive: Being a Compilation of All Acts of a General Nature Now in Force with Full References to Repealed Acts, Special and Local Legislation, and Statutory Constructions of the Supreme Court. San Francisco: H. H. Bancroft and Company.
Hodge, Christina J.
2013 Non-Bodies of Knowledge: Anatomized Remains from the Holden Chapel Collection, Harvard. *Journal of Social Archaeology* 13(1):122–149.
Hodge, Christina J., Jane Lyden Rousseau, and Michèle E. Morgan
2017 Teachings of the Dead: The Archaeology of Anatomized Remains from Holden Chapel, Harvard University. In *The Bioarchaeology of Dissection and Autopsy in the United States,* edited by Kenneth C. Nystrom, pp. 115–142. Springer International, Heidelberg.
Iacumin, Paola, Hervé Bocherens, André Mariotti, and Antonio Longinelli
1996 Oxygen Isotope Analyses of Co-Existing Carbonate and Phosphate in Biogenic Apatite: A Way to Monitor Diagenetic Alteration of Bone Phosphate? *Earth and Planetary Science Letters* 142(1–2):1–6.

Koch, Paul L., Noreen Tuross, and Marilyn L. Fogel
1997 The Effects of Sample Treatment and Diagenesis on the Isotopic Integrity of Carbonate in Biogenic Hydroxyapatite. *Journal of Archaeological Science* 24:417–429.

Lee, Erika
2002 The Chinese Exclusion Example: Race, Immigration, and American Gatekeeping, 1882–1924. *Journal of American Ethnic History* 21(3):36–62.

Luz, Boaz, and Yehoshua Kolodny
1989 Oxygen Isotope Variation in Bone Phosphate. *Applied Geochemistry* 4:317–323.

Manolagas, Stavros
2000 Birth and Death of Bone Cells: Basic Regulatory Mechanisms and Implications for the Pathogenesis and Treatment of Osteoporosis. *Endocrine Review* 21:115–137.

Merriam, Paul G.
1979 The "Other Portland": A Statistical Note on Foreign-Born, 1860–1910. *Oregon Historical Quarterly* 80(3): 258–268.

Meyer, A. W.
1930 Anatomy Acts of California: A Survey of Former and Present Laws. *California and Western Medicine* 33(4):703–707.

Mosier, Page
2001 A Brief History of Population Growth in the Greater San Francisco Bay Region. *UG Survey, Geology and Natural History of the San Francisco Bay Area: A.*

Nystrom, Kenneth C.
2017 Skeletal Evidence of Anatomical and Surgical Training in Nineteenth-Century Richmond. In *The Bioarchaeology of Dissection and Autopsy in the United States*, edited by Kenneth C. Nystrom, pp. 143–164. Springer International Publishing, Heidelberg.

Pate, F. Donald
1994 Bone Chemistry and Paleodiet. *Journal of Archaeological Method and Theory* 1(2):161–209.

Sappol, Michael
2002 *A Traffic of Dead Bodies: Anatomy and Embodied Social Identity in Nineteenth-Century America.* Princeton University Press, Princeton, New Jersey.

tenBroek, Jacobus
1957 California's Welfare Law—Origins and Development. *California Law Review* 45(3):241–303.

Walker, Richard
2001 Industry Builds the City: The Suburbanization of Manufacturing in the San Francisco Bay Area, 1850–1940. *Journal of Historical Geography* 27(1):36–57.

Wollenberg, Charles
1985 *Golden Gate Metropolis: Perspectives on Bay Area History.* Institute of Governmental Studies, Berkeley.

13

Diet Dissected

Isotopic Variation at Point San Jose

SARAH A. HALL, JULIA PRINCE-BUITENHUYS, AND ERIC J. BARTELINK

The pit behind the Point San Jose hospital at Fort Mason, San Francisco, dating to 1860–1900, held approximately 4,450 commingled elements from at least 25 individuals. The bones were recovered along with broken bottles and faunal remains from the pit feature, with several human elements showing evidence of anatomical dissection. In the late nineteenth century, dissection carried a strong negative stigma and was therefore practiced predominately using the bodies of poor individuals with little institutional power or protection (Garment et al. 2007; Humphrey 1973; Sappol 2002; see Milligan, this volume). Various lines of evidence reported in this volume suggest that some of the individuals interred in the pit were likely not born in the San Francisco Bay Area, of low socioeconomic status, and probably include individuals of both European and non-European ancestry (Bartelink and Hall, this volume; Sgheiza and Willey, this volume).

Given the diverse backgrounds of the individuals interred at Point San Jose in terms of demography and region of origin, in this chapter we evaluate stable isotope evidence to explore the extent to which foodways reflected diverse group identities among these marginalized individuals. Dietary customs can illuminate aspects of community health and social identity, including ethnicity, socioeconomic status, and religion. Thus, stable isotope analysis (SIA) of human bone can be a powerful tool for reconstructing foodways and life histories of individuals. This study uses stable carbon and nitrogen isotope analysis of human bone to reconstruct the diet of individuals interred in the Point San Jose pit. The Santa Clara Valley Medical Center (SCVMC), a partially contemporaneous pauper cemetery

(1875–1935) located 42 miles to the south in Santa Clara County, California, serves as a local comparative sample. In addition, comparative data from a multiregional US sample of varied socioeconomic status is used to evaluate the effects of regional and socioeconomic influences on foodways of those interred at Point San Jose.

Foodways of Individuals Interred in the Pit at Point San Jose

Bioarchaeological studies commonly examine changing foodways to evaluate ecological, political, and social changes in the past. Foodways encompass more than dietary habits; they "constitute an organized system, a language that—through its structure and components—conveys meaning and contributes to the organization of the natural and social world" (Counihan 2018 [1999]:19). As Jean Brillat-Savarin (2009 [1825]) aptly put it, "Tell me what you eat, and I shall tell you what you are." Anthropological studies have long documented dietary practices, including tastes and taboos (e.g., Mallery 1888), but studies over the last half century have explicitly theorized the role of foodways (Counihan and van Esterik 2013), acknowledging the importance of dietary practices in forming and performing group identities (Douglas 1984; Harris 1985; Levi-Strauss 1963) and in establishing social status (Bourdieu 1984; Goody 1982; Mintz 1979, 1985). Foodways are dynamic and shift to accommodate environmental availability of foods as well as cultural practices, including technological advancements and social connotations of food practices; consequently, cultural changes often result in new foodways (Dietler 2007; Mintz and Du Bois 2002).

Changing dietary patterns are used to infer shifts in subsistence strategies, sociopolitical dynamics, and population health among communities in the past and present. These dietary differences can occur along numerous axes of identity. Contemporary studies show that migrant populations often maintain aspects of their native foodways rather than adopting those of the country where they moved (Brown and Mussell 1984; Douglas 1984). This retention often symbolizes and signals group identity (Kalčik 1984; Murcott 1996), although it can also serve to stigmatize those same groups (Abrahams 1984; Harbottle 1997). Archaeological studies have explored the construction of group identities through shared food practices (Hastorf 2016; Twiss 2007). Many studies have employed stable isotope analysis to evaluate dietary differences based on status (Ambrose et al. 2003; Reitsema and Vercellotti 2012; Somerville et al. 2013) and dietary assimilation after migration (Laffoon et al. 2018; White et al. 2004). In regard to the Point San Jose assemblage, the dietary components of migrant foodways and differences that occurred along distinctions of socioeconomic status are of particular interest. Importantly, this study uses diet as a proxy for foodways with the understanding that diet is only one element of cultural food practices. Additionally, migration is defined broadly as a change in residence across substantial cultural, political, and/or environmental boundaries (Tsuda et al. 2015) to include foreign-born and US-born individuals who moved to the Bay Area.

In the late 1800s when the Point San Jose pit feature was created, the most common migrants to San Francisco were mostly white US citizens as well as foreign-born individuals from China, Japan, Central and South America, and Europe, who rapidly displaced Californios and Native Americans (Almaguer 1994; Barkan 2007; Berglund 2010; Burchell 1980; Chen 2002; Daniels 1990). According to the 1870 US Census (1872), San Francisco had a "White" majority (>90%). Based on the categories reported in the census, approximately 8 percent of the population was "Chinese" (mainly from the Guangdong Province, followed by Guangxi and Fujian Provinces; Farkas 1998; Voss et al. 2018), 0.8 percent was "Free Colored," and 0.03 percent was "Indian" (US Census Bureau 1872). Although undoubtedly a major destination for foreign immigration, a slight majority (50.7%) of individuals was US-born (US Census Bureau 1872). It is important to note, however, that census reports from this period had a tendency to undercount some groups, resulting in under-enumeration of immigrants, poor, and/or indigent persons (Reid 1995). Likewise, the recording and conceptualization of race and racial categories was not the same as it is in the modern day (Snipp 2003).

By the late 1800s, preservation and transportation technologies permitted routine import and export of foods in the American West (Gabaccia 1998). This mobility was particularly important for San Francisco, which relied heavily on imports from across California as well as national and international agricultural trade (Peters 2013). Consequently, a variety of foodstuffs was available in San Francisco. This trade suggests it was possible that migrants to California were able to maintain their original foodways without adopting a locally sourced diet, provided that regional and socioeconomic influences on migrant foodways were not limiting. Furthermore, trade across the continent made it possible for individuals of a given socioeconomic status to share a similar diet despite regional differences. It is also possible that regional influences dominated migrant foodways, limiting the effects of transcontinental trade on migrant diets in San Francisco.

Stable Isotope Analysis

SIA is often used to infer patterns of diet and migration in past populations. This study evaluates the effects of socioeconomic status and geographic region on diet, measured by evaluating variation in carbon (from protein and non-protein sources) and nitrogen (from dietary protein) using two components of human bone: collagen and bioapatite.

Human bone collagen and bone bioapatite reflect different contributions of dietary macronutrients (Ambrose and Norr 1993; Krueger and Sullivan 1984; Tieszen and Fagre 1993). Bone collagen preferentially incorporates carbon from protein sources because essential amino acids come from dietary protein. Carbon isotopes of bioapatite carbonate derive from dissolved bicarbonate in the blood, which comes from dietary carbohydrates, lipids, and proteins not used in tissue synthesis. Thus, bioapatite reflects the whole diet. Because collagen and bioapatite reflect different dietary components, it is useful to consider both constituents for dietary reconstructions (e.g., Ambrose et al. 2003; Froehle et al. 2010; Harrison and Katzenberg 2003).

This study focuses on two stable isotopes that reflect diet: carbon ($^{13}C/^{12}C$) and nitrogen ($^{15}N/^{14}N$). Carbon isotopes reflect the photosynthetic pathways plants use to fix carbon (DeNiro and Epstein 1978). The food sources relevant to this study rely on two different photosynthetic pathways: C_3, used by temperate-region vegetation including barley, fruit, legumes, rice, tubers, and wheat; and C_4, used by tropical and subtropical vegetation including cane sugar and maize/corn (Heaton 1999; O'Leary 1988; Smith and Epstein 1971). C_3 plants are more depleted in ^{13}C than C_4 plants, and consequently these two pathways produce distinct, non-overlapping ranges of isotopic values (van der Merwe 1982). Nitrogen isotopes reflect the trophic level of an organism, such that the $\delta^{15}N$ value of that organism is elevated 3–4‰ over its diet (DeNiro and Epstein 1981; Minagawa and Wada 1984; Schoeninger and DeNiro 1984).

Marine carbon values appear distinct from terrestrial sources, but frequently overlap with C_4 resources (Schoeninger and DeNiro 1984). However, because food chains are longer in marine ecosystems, nitrogen isotope values are more elevated, thus the combination of carbon and nitrogen values can differentiate marine from terrestrial food sources (Schoeninger 1995; Schoeninger and DeNiro 1984).

For the nineteenth-century Bay Area, most plant sources were C_3-based; wheat and fruit were the two main agricultural commodities (Jelinek 1979). However, by this period, US ranchers also used corn and other C_4 grains to feed cattle (Gabaccia 1998), so it is likely that the local diet would have included both C_3 and C_4 sources. Local protein sources would have included beef (another major industry in the region) and freshwater and marine fish and shellfish.

Although the Bay Area would have provided a local diet based on C_3 and C_4 carbon sources and terrestrial, marine, and freshwater protein sources, it should not be assumed that residents shared a similar diet. Isotopically, dietary variability among migrants from the eastern US, Europe, East Asia, and Latin America might differ widely. Contemporary studies suggest individuals from many parts of Latin America have very high carbon isotope values due to their heavy reliance on corn (a C_4 plant) and corn products. On the other end of the spectrum, East Asians and Europeans generally have very low carbon isotope values, reflecting greater C_3 resource consumption and a lack of C_4 resources in the diet; US diets typically fall in between these extremes (Bartelink et al. 2014, 2018; Hülsemann et al. 2015).

Materials and Methods

Human bone samples from the Point San Jose pit were obtained from 22 right tali (the most common skeletal element from the assemblage) and 8 skulls for a total sample size of 30. Comparative samples from eighteenth- to nineteenth-century North America include 50 human samples from SCVMC (Prince-Buitenhuys 2016) and 97 human bone samples from other US burial sites (France et al. 2014). All sampled individuals were adults. Individuals were categorized by region and socioeconomic status based on site location and historical records, following the designations used by France and colleagues (2014). The Point San Jose and SCVMC samples were considered as separate regional categories from France et al. (2014) samples. The SCVMC sample was used only as a regional comparison.

Bone collagen was extracted following Ambrose (1990). Structural carbonate of bioapatite was extracted according to Koch et al. (1997). Collagen samples were analyzed at the University of California, Davis, Department of Plant Sciences Stable Isotope Facility. Values for $\delta^{13}C$ and $\delta^{15}N$ were collected using a PDZ Europa ANCA-GSL elemental analyzer with a PDX Europa 20–20 isotope ratio mass spectrometer (internal precision of ± 0.2‰ for $\delta^{13}C$ and ± 0.3‰ for $\delta^{15}N$). Bioapatite samples were analyzed by the Stable Isotope Facility at the University of Wyoming. $\delta^{13}C$ was measured using a Thermo Gasbench coupled with a Thermo Delta Plus XL continuous-flow inlet isotope ratio mass spectrometer with an internal precision of ± 0.3‰ for $\delta^{13}C$. Samples with collagen yields above 1 percent and atomic C/N ratios within the established 2.9 to 3.6 range were deemed acceptable (Ambrose 1990; DeNiro 1985; van Klinken 1999).

Descriptive statistics were generated based on region and estimated socioeconomic status. Inter-group comparisons among regions and socioeconomic statuses were conducted using nonparametric Kruskal-Wallis and post hoc Dunn tests. A within-group analysis for the Point San Jose sample was conducted using compiled historical Bay Area foodweb data (see Prince-Buitenhuys 2016). Finally, carbon isotope modeling (Froehle et al. 2010) was used to examine the relationship between carbon values of bioapatite versus collagen $\delta^{13}C$. Because these two tissues record different dietary inputs, bivariate modeling helps to clarify the contributions of different dietary macronutrients. Carbon values were adjusted by –1.5‰ to use the Froehle et al. (2010) model, because the model was created using modern tissues (post-Seuss Effect). Statistical analyses were conducted using R and SPSS v.24.

Results and Discussion

Regional Comparisons

Table 13.1 reports the descriptive statistics and statistical results for the Point San Jose sample and the three comparative samples. The $\delta^{13}C$ and $\delta^{15}N$ values were significantly different between Point San Jose and all other regions ($p < 0.05$) except SCVMC. These results show that individuals interred in the pit at Point San Jose had the lowest collagen $\delta^{13}C$ values and highest $\delta^{15}N$ values of the four samples.

Socioeconomic Status Comparisons

Table 13.2 reports the descriptive statistics and statistical results for the Point San Jose sample and the three comparative samples by socioeconomic status (SES). $\delta^{13}C$ collagen values from Point San Jose differed significantly from all SES groups ($p < 0.05$). $\delta^{15}N$ values from Point San Jose were significantly different ($p < 0.05$) from all SES groups except the Mid-

Table 13.1. Descriptive and inferential statistics for Point San Jose (PSJ) and region-of-origin using carbon and nitrogen values (‰) from collagen

		$\delta^{13}C$ Descriptive Statistics		$\delta^{13}C$ Comparison with PSJ			$\delta^{15}N$ Descriptive Statistics		$\delta^{15}N$ Comparison with PSJ		
Region of Origin	N	Mean	Std. Dev	Test Statistic	Std. Error	Adj. Sig.	Mean	Std. Dev.	Test Statistic	Std. Error	Adj. Sig.
SCVMC	50	−17.1	1.1	21.323	11.833	0.429	11.6	0.8	-26.283	11.831	0.158
Northern US	72	−15.4	2.2	61.724	11.134	0.000*	11.0	0.8	-63.169	11.132	0.000*
Southern US	25	−13.1	1.6	102.673	13.875	0.000*	11.0	1.0	-60.623	13.873	0.000*
Point San Jose	30	−17.8	1.8				12.0	0.7			

* Statistically significant comparisons.

Table 13.2. Descriptive and inferential statistics for Point San Jose (PSJ) and socioeconomic status groups using carbon and nitrogen values (‰) from collagen

		$\delta^{13}C$ Descriptive Statistics		$\delta^{13}C$ Comparison with PSJ			$\delta^{15}N$ Descriptive Statistics		$\delta^{15}N$ Comparison with PSJ		
Socioeconomic Status	N	Mean	Std. Dev	Test Statistic	Std. Error	Adj. Sig.	Mean	Std. Dev.	Test Statistic	Std. Error	Adj. Sig.
Lower Class	9	−14.3	2.2	57.722	13.987	0.000*	10.3	0.71	-67.722	13.982	0.000*
Middle Class	6	−12.3	1.1	81.333	16.458	0.000*	11.2	1.00	-33.583	16.453	0.412
Military	43	−14.6	2.6	47.946	8.754	0.000*	11.0	0.85	-41.267	8.752	0.000*
Upper Class	39	−15.5	1.7	39.077	8.937	0.000*	11.1	0.80	-41.167	8.934	0.000*
Point San Jose	30	−17.8	1.8				12.0	0.69			

* Statistically significant comparison.

dle Class sample. These results show that individuals at Point San Jose had relatively lower collagen $\delta^{13}C$ values and relatively higher $\delta^{15}N$ values than most other socioeconomic groups.

Dietary Variation at Point San Jose

Analyses by region demonstrated that the carbon and nitrogen isotope values from the Point San Jose sample were not significantly different from the other California sample (SCVMC) but were significantly different from other regions. Therefore, it is appropriate to plot the Point San Jose data against a historical Bay Area foodweb (Figure 13.1).

Figure 13.1 shows that individuals interred in the pit at Point San Jose consumed terrestrial animals and animal byproducts as their primary source of protein, although values could also be consistent with some marine and freshwater fish consumption. However, $\delta^{13}C$ values indicate a more C_3–based diet. C_4 and/or marine resources were less important but did contribute to the overall diet for most individuals.

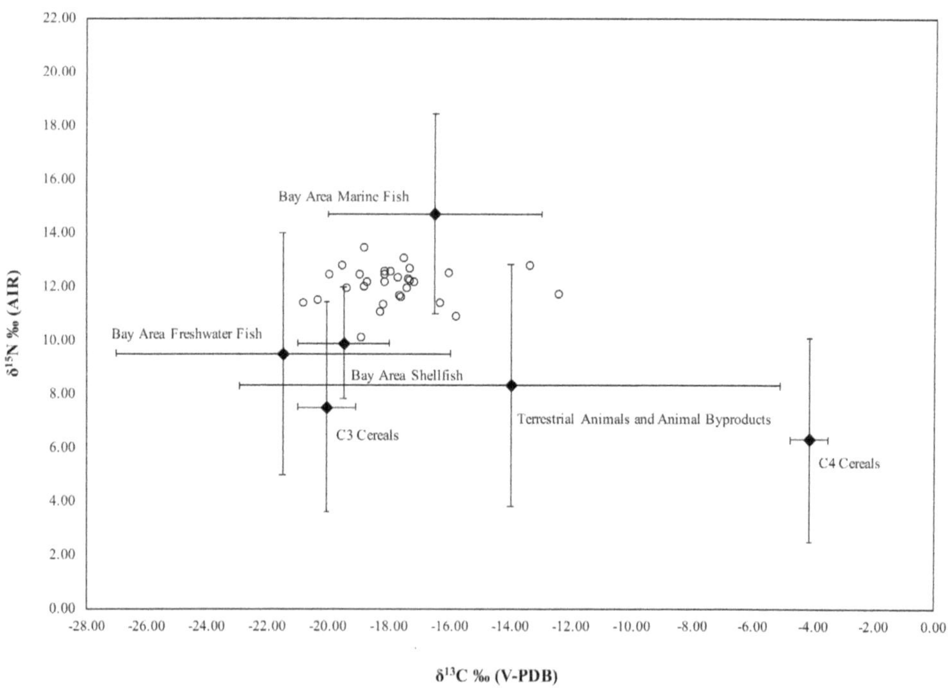

Figure 13.1. Bivariate plot of $\delta^{13}C$ and $\delta^{15}N$ values from Point San Jose bone collagen values and historical Bay Area foodweb (based on Prince-Buitenhuys 2016), California during nineteenth–twentieth centuries.

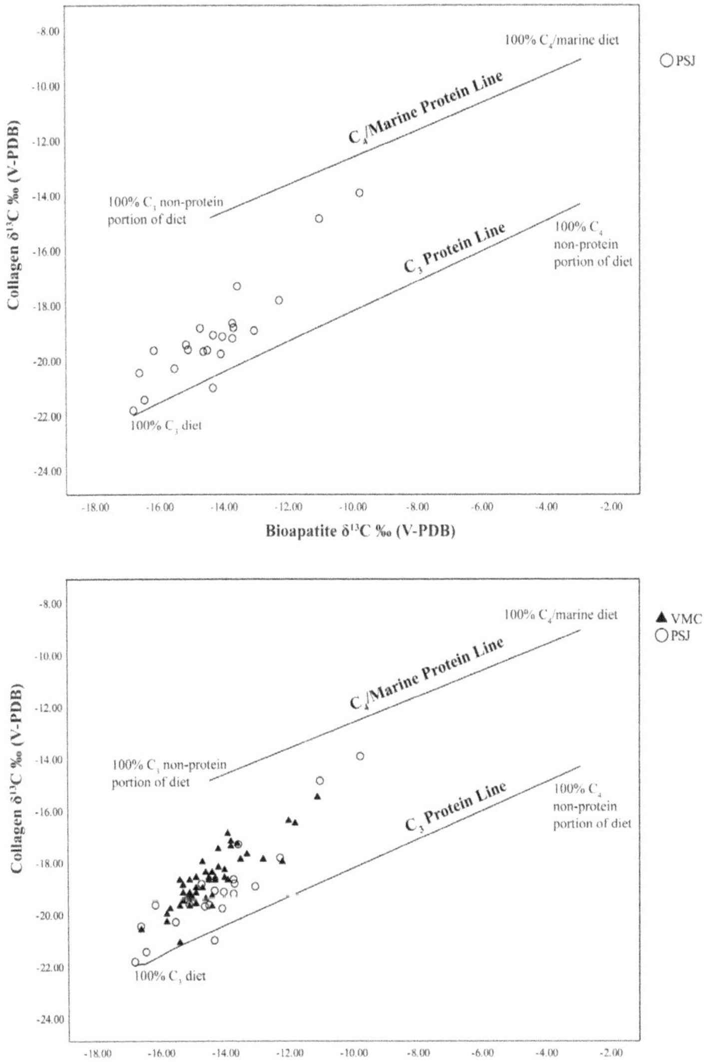

Figure 13.2. Bivariate plots of $\delta^{13}C$ from bioapatite and $\delta^{13}C$ from collagen for Point San Jose (PSJ) and Santa Clara Valley Medical Center (VMC). *Above*: Point San Jose. *Below*: Both Point San Jose and Santa Clara Valley Medical Center samples.

Carbon isotope values of bone collagen and bioapatite were plotted against the Froehle et al. (2010) carbon isotope model in Figure 13.2. This model indicates that most individuals relied on C_3-based foods for the non-protein component of their diet. Additionally, the model suggests that the protein component of the diet for the Point San Jose individuals varied widely, deriving from a combination of C_3 and C_4 and/or marine foods. Finally, comparison of the Point San Jose sample with the SCVMC sample indicates that individuals from both locations consumed isotopically similar food resources.

Implications for Diet of the Dissected

Carbon and nitrogen values at Point San Jose were significantly different from all regions except the SCVMC sample from California. Dietary patterns at Point San Jose and SCVMC were heavily influenced by local agriculture, in apparent contradiction to contemporary studies, which have shown that migrants often maintain certain aspects of their foodways (Brown and Mussell 1984; Douglas 1984). However, SIA does not reflect cuisine, only raw dietary components; migrants may have maintained foodways through other means, such as food preparation techniques. Maintaining traditional food preparation practices has been supported in archaeological and ethnographic studies of Chinese and Japanese Americans in California (Cummings et al. 2014; Matsumoto 1993; Praetzellis and Praetzellis 1982). Although multiple ancestral groups are likely represented in the Point San Jose collection and stable strontium and oxygen isotope data suggest that many individuals were of nonlocal origin (Bartelink and Hall, this volume), the diets of individuals at Point San Jose were consistent with consumption of local foodstuffs. Potential explanations for these results are discussed below.

Results also indicated that on a national scale, regional patterning superseded the effects of SES for the Point San Jose collection. Although these individuals were likely low SES, $\delta^{13}C$ values at Point San Jose were significantly different from all other SES groups (including other individuals of low SES) and $\delta^{15}N$ values were significantly different from all groups except the Middle Class sample. Previous studies have shown local factors, including local socioeconomic and sociopolitical dynamics, tend to influence diet more than broad regional or national trends (Nado 2017; Reitsema et al. 2017). The influence of SES at Point San Jose should therefore not be dismissed but contextualized regionally. For example, a southern burial ground for enslaved persons was classified as "Lower Class," but it is unlikely that this group experienced low SES the same way as did individuals at Point San Jose. Although both were marginalized, enslaved Black men and women and poor Bay Area individuals reflect vastly different life experiences and foodways. Thus, future analyses should incorporate local Bay Area individuals of other social classes for comparison.

Individuals interred in the pit at Point San Jose did not consume isotopically diverse food resources, but there are limitations to the assumption that diverse foodways are reflected as isotopic heterogeneity. It is possible the native foodways for all of the individuals in the Point San Jose sample were isotopically similar to Bay Area resources, and thus there was no isotopic difference between local dietary resources and their traditional cuisine. Furthermore, although the Point San Jose sample demonstrates isotopic homogeneity and is comparable to the SCVMC sample, it is possible for relatively diverse foodways to exist within the same general diet. Many foreign immigrants to the Bay Area, for example, came from countries where traditional cuisines were based on C_3 resources, and consumed both terrestrial animals and seafood, making their traditional diets isotopically similar with the resources available in the Bay Area (Peters 2013). Nonetheless, based on data from the comparative US sample (France et al. 2014), it is clear other US regions were distinct from the Bay Area diet.

Alternatively, if Point San Jose included individuals with traditional diets dissimilar from resources in the Bay Area, such as those from the eastern US (France et al. 2014) or even Californios (Peters 2013), then other factors like cost or stigma may have prevented the incorporation of traditional foods into their diets, despite their availability as a result of transcontinental and international trade in the American West (Gabaccia 1998). It is worth noting that even among foodways compatible with Bay Area resources, cost may have been a limiting factor in an individual's ability to maintain their group identity through foodways. For example, many Chinese workers could not regularly afford to eat traditional fare in Chinatown, only occasionally affording a meal from a street vendor or grocer (Peters 2013:51).

This study also does not address diet quality. Once again, it is important to avoid sweeping generalities about the "sameness" of diet in a given region across diverse groups (Brown and Mussell 1984; Douglas 1984). These generalizations often elide important nutritional differences (e.g., Joffe 1943; Whitehead 1992). This caveat reinforces the need for a larger sample of Bay Area individuals, both migrant and local, of diverse backgrounds and across socioeconomic classes to improve interpretations of nineteenth-century Bay Area foodways.

Overall, the isotope data presented here provide support for strong regional influences on Point San Jose foodways. Based on the relatively low $\delta^{13}C$ values and relatively high $\delta^{15}N$ values observed in both of the samples at Point San Jose and at SCVMC, the California Bay Area diet in the nineteenth century was dominated by C_3 plant sources and higher trophic-level protein sources. Because the values of the Point San Jose sample were consistent with those from the SCVMC sample, it was appropriate to compare Point San Jose samples with a historical Bay Area foodweb. The Froehle et al. (2010) model depicted in Figure 13.2 illustrates the relationship between $\delta^{13}C$ from the protein and non-protein components of the diet. The non-protein portion of the diet relied heavily on C_3 plants; based on agriculture in the area at the time, probably wheat and fruit (Jelinek 1979). The protein portion of the diet at Point San Jose was also dominated by C_3 sources. The relatively low $\delta^{13}C$ and high $\delta^{15}N$ values are consistent with values from terrestrial animals and freshwater fish, although a small contribution from C_4 or marine food sources cannot be excluded. During this period, corn was a key resource particularly for feed grain (Gabaccia 1998), and likely contributed to elevated $\delta^{13}C$ values in the eastern US comparative samples. However, corn was not a major crop in northern California (Jelinek 1979; Rice et al. 2002). While it is possible that the C_4 inputs in the protein component of the diet came from corn-based diets for terrestrial animals, it is more likely that individuals at Point San Jose were consuming fish resources considering the geography and economy of San Francisco (Peters 2013).

Limitations

Several factors limit interpretation of these data, including commingling at Point San Jose, small sample size, lack of California-specific comparative data for socioeconomic status, and lack of faunal isotope values for baseline data. First of all, the problems of commingling and small sample sizes are related. Because the assemblage was commingled, isotope sampling was limited to the most common skeletal element to avoid sampling a single individual more than once. This approach limited the number of isotopic samples taken from the Point San Jose sample. Additionally, some categories of SES in the comparative sample were small.

It was expected that Point San Jose sample would be significantly different from other low SES samples considering the documented effect of local foodways on dietary patterns. Nonetheless, studies have also shown that local status differences can impact diet. Therefore, future studies should incorporate comparative California samples from Upper and Middle Class individuals to evaluate the effects of local socioeconomic dynamics on foodways. Additionally, faunal isotopic data from the same period would permit a more fine-grained understanding of the human isotope values.

Summary

This study indicates that regional dietary patterns dominated Point San Jose foodways in the Bay Area in the nineteenth century. Although transcontinental trade made diverse foodways possible, the isotopic data are relatively homogeneous. Further, comparisons with contemporaneous samples across the nation suggest that socioeconomic influences did not create dietary differences by social class, at least insofar as this is reflected by isotopic diversity.

Analyses of carbon and nitrogen isotopes among individuals interred in the pit at Point San Jose indicated relatively homogeneous values, suggesting a similar dietary emphasis. These individuals showed relatively high nitrogen isotope values and relatively low carbon isotope values, indicating a dietary emphasis on C_3 plants, terrestrial animals, and freshwater fish. Individuals from the Point San Jose assemblage shared a diet similar to that of other contemporaneous individuals in the region of similar socioeconomic status.

The individuals interred in the pit at Point San Jose did not share a diet with other contemporaneous individuals of low SES from other regions. Although it is clear that on a national level, regional patterns had a stronger influence on diet than did SES, these data did not permit evaluations of the effects of *local* socioeconomic and sociopolitical impacts on migrant dietary patterns at Point San Jose. This approach requires further research. With a larger sample from the Bay Area that includes migrants and locals across social classes, the impacts of SES on foodways may be better evaluated.

Finally, isotopic data at Point San Jose merely reflect the types of foods consumed by individuals from this site and can oversimplify complex foodways. Many studies have shown that nuances of group identity are reflected not only in foods consumed, but in cultural practices surrounding food preparation, fellowship, and food consumption (Goode et al. 1984; James 1997; Williams-Forson 2008). Cuisine is about rules (Fischler 1988). Adherence to those rules, rather than the foods they allow or proscribe, is often used to delineate group identity and status (Appadurai 1988; James 1997). This study demonstrates that individuals buried at Point San Jose had diets composed of the same foods; whether those diets followed the same "rules" is another question entirely and begs further research.

References Cited

Abrahams, Roger

1984 Equal Opportunity Eating: A Structural Excursus on Things of the Mouth. In *Ethnic and Regional Foodways in the United States: The Performance of Group Identity,* edited by L. K. Brown and K. Mussell, pp. 1–36. University of Tennessee Press, Knoxville.

Almaguer, Tomas

1994 *Racial Fault Lines: The Historical Origins of White Supremacy in California.* University of California Press, Berkeley.

Ambrose, Stanley H.

1990 Preparation and Characterization of Bone and Tooth Collagen for Isotope Analysis. *Journal of Archaeological Sciences* 17:431–451.

Ambrose, Stanley H., Jane E. Buikstra, and H. W. Krueger

2003 Status and Gender Differences in Diet at Mound 72, Cahokia, Revealed by Isotopic Analysis of Bone. *Journal of Anthropological Archaeology* 22:217–226.

Ambrose, Stanley H., Lynette C. Norr

1993 Experimental Evidence for the Relationship of the Carbon Isotope Ratios of Whole Diet and Dietary Protein to Those of Bone Collagen and Carbonate. In *Prehistoric Human Bone: Archaeology at the Molecular Level,* edited by J. B. Lambert and G. Grupe, pp. 1–37. Springer-Verlag, Berlin.

Appadurai, Arjun

1988 How to Make a National Cuisine: Cookbooks in Contemporary India. *Comparative Studies in Society and History* 30(1):3–24.

Barkan, Elliott R.

2007 *From All Points: America's Immigrant West 1870s–1952.* Indiana University Press, Bloomington.

Bartelink, Eric J., Gregory E. Berg, Melanie M. Beasley, and Lesley A. Chesson

2014 Application of Stable Isotope Forensics for Predicting Region of Origin of Human Remains from Past Wars and Conflicts: Human Remains from Past Wars and Conflicts. *Annals of Anthropological Practice* 38(1):124–136.

Bartelink, Eric J., Gregory E. Berg, Lesley A. Chesson, Brett J. Tipple, Melanie M. Beasley, Julia R. Prince-Buitenhuys, Heather MacInnes, Amy T. MacKinnon, and Krista E. Latham.

2018 Applications of Stable Isotope Forensics for Geolocating Unidentified Human Remains from Past Conflict Situations and Large-Scale Humanitarian Efforts. In *New Perspectives in Forensic Human Skeletal Identification,* edited by K. E. Latham, E. J. Bartelink, and M. Finnegan, pp. 175–184. Elsevier Academic, San Diego.

Berglund, Barbara

2010 *Making San Francisco American: Cultural Frontiers in the Urban West, 1846–1906.* University Press of Kansas, Lawrence.

Bourdieu, Pierre

1984 *Distinction: A Social Critique of the Judgement of Taste.* Harvard University Press, Cambridge, Massachusetts.

Brillat-Savarin, Jean A.
2009 [1825] *The Physiology of Taste: Or Meditations on Transcendental Gastronomy.* Translated by M.F.K. Fisher. Knopf Doubleday Publishing Group, New York. (Originally published in, translated in 1949).

Brown, Linda K., and Kay Mussell (editors)
1984 *Ethnic and Regional Foodways in the United States: The Performance of Group Identity.* University of Tennessee Press, Knoxville.

Burchell, Robert A.
1980 *The San Francisco Irish, 1848–1880.* University of California Press, Berkeley.

Chen, Yong
2002 *Chinese San Francisco, 1850–1943: A Trans-Pacific Community.* Stanford University Press, Stanford, California.

Counihan, Carole M.
2018 [1999] Food, Culture, and Gender. In *The Anthropology of Food and Body: Gender, Meaning, and Power,* edited by C. M. Counihan, pp. 6–24. Routledge, New York.

Counihan, Carole, and Penny Van Esterik (editors).
2013 *Food and Culture: A Reader.* Routledge.

Cummings, Linda Scott, Barbara L. Voss, Connie Young Yu, Peter Kovacik, Kathryn Puseman, Chad Yost, Ryan Kennedy, and Megan S. Kane
2014 Fan and Tsai: Intracommunity Variation in Plant-Based Food Consumption at the Market Street Chinatown, San José, California. *Historical Archaeology* 48(2):143–172.

Daniels, Douglas H.
1990 *Pioneer Urbanites: A Social and Cultural History of Black San Francisco.* University of California Press, Berkeley.

DeNiro, Michael J.
1985 Postmortem Preservation and Alteration of *in vivo* Bone Collagen Isotope Ratios in Relation to Palaeodietary Reconstruction. *Nature* 317: 806–809.

DeNiro, Michael J., and Samuel Epstein
1978 Influence of Diet on the Distribution of Carbon Isotopes in Animals. *Geochimica et Cosmochimica Acta* 42:495–506.
1981 Influence of Diet on the Distribution of Nitrogen Isotopes in Animals. *Geochimica et Cosmochimica Acta* 45:341–351.

Dietler, Michael
2007 Culinary Encounters: Food, Identity, and Colonialism. In *The Archaeology of Food and Identity,* edited by K. Twiss, pp. 218–242. Center for Archaeological Investigations Press, Carbondale, Illinois.

Douglas, Mary
1984 *Food in the Social Order.* Russell Sage, New York.

Farkas, Lani Ah Tye
1998 *Bury My Bones in America.* Carl Mautz, Nevada City, California.

Fischler, Claude
1988 Food, Self, Identity. *Social Science Information* 27(2):275–292.

France, Christine A. M., Douglas W. Owsley, and Lee-Ann C. Hayek

2014 Stable Isotope Indicators of Provenance and Demographics in 18th and 19th Century North Americans. *Journal of Archaeological Sciences* 42:356–366.

Froehle, Andrew W., Corina M. Kellner, and Margaret J. Schoeninger
2010 FOCUS: Effect of Diet and Protein Source on Carbon Stable Isotope Ratios in Collagen: Follow Up to Warinner and Tuross 2009. *Journal of Archaeological Sciences* 37:2662–2670.

Gabaccia, Donna R.
1998 *We Are What We Eat: Ethnic Food and the Making of Americans.* Harvard University Press, Cambridge, Massachusetts.

Garment, Ann, Susan Lederer, Naomi Rogers, and Lisa Boult
2007 Let the Dead Teach the Living : The Rise of Body Bequeathal in 20th-Century America. *Academic Medicine* 82(10):1000–1005.

Goode, Judith, Karen Curtis, and Janet Theophano
1984 Meal Formats, Meal Cycles, and Menu Negotiation in the Maintenance of an Italian American Community. In *Food in the Social Order,* edited by Mary Douglas, pp. 143–218. Russell Sage, New York.

Goody, Jack
1982 *Cooking, Cuisine, and Class: A Study in Comparative Sociology.* Cambridge University Press, Cambridge, United Kingdom.

Harbottle, Lynn
1997 Fast Food/Spoiled Identity: Iranian Migrants in the British Catering Trade. In *Food, Health, and Identity,* edited by P. Caplan, pp. 87–110. Routledge, London.

Harris, Marvin
1985 *Good to Eat: Riddles of Food and Culture.* Simon and Schuster, New York.

Harrison, Roman G., and M. Anne Katzenberg
2003 Paleodiet Studies Using Stable Carbon Isotopes from Bone Apatite and Collagen: Examples from Southern Ontario and San Nicolas Island, California. *Journal of Anthropological Archaeology* 22:227–244.

Hastorf, Christine A.
2016 *The Social Archaeology of Food: Thinking about Eating from Prehistory to the Present.* Cambridge University Press.

Heaton, Tim H. E.
1999 Spatial, Species, and Temporal Variations in the $^{13}C/^{12}C$ Ratios of C_3 Plants: Implications for Paleodiet Studies. *Journal of Archaeological Sciences* 26:637–649.

Hülsemann, Frank, Christine Lehn, Sabine Schneiders, Glen Jackson, Sarah Hill, Andreas Rossmann, . . . Wilhelm Schänzer
2015 Global Spatial Distributions of Nitrogen and Carbon Stable Isotope Ratios of Modern Human Hair. *Rapid Communications in Mass Spectrometry* 29(22):2111–2121.

Humphrey, David C.
1973 Dissection and Discrimination: The Social Origins of Cadavers in America, 1760–1915. *Bulletin of the New York Academy of Medicine* 49:819–825.

James, Allison
1997 How British is British Food? In *Food, Health, and Identity,* edited by Pat Caplan, pp. 71–86. Routledge, London.

Jelinek, Lawrence J.
1979 *Harvest Empire: A History of California Agriculture.* Boyd and Fraser, San Francisco.

Joffe, Natalie F.
1943 Food Habits of Selected Subcultures in the United States. *Bulletin of National Research Council* 108:98.

Kalčik, Susan
1984 Ethnic Foodways in America: Symbol and the Performance of Identity. In *Ethnic and Regional Foodways in the United States: The Performance of Group Identity,* edited by L. K. Brown and K. Mussel, pp. 37–65. University of Tennessee Press, Knoxville.

Koch, Paul L., Noreen Tuross, and Marilyn L. Fogel
1997 The Effects of Sample Treatment and Diagenesis on the Isotopic Integrity of Carbonate in Biogenic Hydroxylapatite. *Journal of Archaeological Sciences* 24:417–429.

Krueger, Harold W., and Charles H. Sullivan
1984 Models for Carbon Isotope Fractionation between Diet and Bone. In *Stable Isotopes in Nutrition,* edited by J. E. Turnlund and P. E. Johnson, pp. 205–222. American Chemical Society Symposium Series 258. Washington, DC.

Laffoon, Jason E., Ryan Espersen, and Hayley L. Mickleburgh
2018 The Life History of an Enslaved African: Multiple Isotope Evidence for Forced Childhood Migration from Africa to the Caribbean and Associated Dietary Change. *Archaeometry* 60(2): 350–365.

Levi-Strauss, Claude
1963 *Totemism.* Beacon Press, Boston, Massachusetts.

Mallery, Garrick
1888 Manners and Meals. *American Anthropologist* 1(3):193–207.

Matsumoto, Valerie J.
1993 *Farming the Home Place: A Japanese American Community in California, 1919–1982.* Cornell University Press, Ithaca, New York.

Minagawa, Masao, and Eitaro Wada
1984 Stepwise Enrichment of ^{15}N along Food Chains: Further Evidence and the Relation between $\delta^{15}N$ and Animal Age. *Geochimica et Cosmochimica Acta* 48:1135–1140.

Mintz, Sidney W.
1979 Time, Sugar, and Sweetness. *Marxist Perspectives* 2(4):56–73.
1985 *Sweetness and Power: The Place of Sugar in Modern History.* Penguin Books, New York.

Mintz, Sidney W., and Christine M. Du Bois
2002 The Anthropology of Food and Eating. *Annual Review of Anthropology* 31:99–119.

Murcott, Anne
1996 Food as an Expression of National Identity. In *The Future of the Nation-State: Essays on Cultural Pluralism and Political Integration,* edited by S. Gustavsson and L. Lewin, pp. 59–96. Routledge, London.

Nado, Kristen

2017 Dietary Practices, Socioeconomic Status, and Social Mobility at Teotihuacan, Mexico. PhD dissertation, Department of Anthropology, Arizona State University, Tempe.

O'Leary, Marion H.

1988 Carbon Isotopes in Photosynthesis. *BioScience* 38:328–336.

Peters, Erica J.

2013 *San Francisco: A Food Biography.* Rowman & Littlefield, Lanham, Maryland.

Praetzellis, Mary, and Adrian Praetzellis

1982 Archaeological and Historical Studies of the IJ56 Block, Sacramento, California: An Early Chinese Community. Sonoma State University Anthropological Studies Center Cultural Resources Facility, Sonoma, California.

Prince-Buitenhuys, Julia R.

2016 New Foods, New Lands: Ethnic Dietary Variation at the Santa Clara Valley Medical Center Potter's Field. Master's thesis, Department of Anthropology, California State University, Chico.

Reid, Richard

1995 The 1870 United States Census and Black Undernumeration: A Test Case from North Carolina. *Histoire Sociale,* 28(56):487–499.

Reitsema, Laurie J., Tomasz Kozłowski, Douglas E. Crews, M. Anne Katzenberg, and Wojciech Chudziak

2017 Resilience and Local Dietary Adaptation in Rural Poland, 1000–1400 CE. *Journal of Anthropological Archaeology* 45:38–52.

Reitsema, Laurie J., and Giuseppe Vercellotti.

2012 Stable Isotope Evidence for Sex-and Status-Based Variations in Diet and Life History at Medieval Trino Vercellese, Italy. *American Journal of Physical Anthropology* 148(4): 589–600.

Rice, Richard B., William A. Bullough, and Richard J. Orsi

2002 *The Elusive Eden: A New History of California* (3rd ed). McGraw-Hill, Boston.

Sappol, Michael

2002 *A Traffic of Dead Bodies: Anatomy and Embodied Social Identity in Nineteenth-Century America.* Princeton University Press, Princeton, New Jersey.

Schoeninger, Margaret J.

1995 Stable Isotope Studies in Human Evolution. *Evolutionary Anthropology: Issues, News, and Reviews* 4(3):83–98.

Schoeninger, Margaret J., and Michael J. DeNiro

1984 Nitrogen and Carbon Isotopic Composition of Bone Collagen from Marine and Terrestrial Animals. *Geochemica et Cosmochemica Acta* 48:625–639.

Smith, Bruce N., and Samuel Epstein

1971 Two Categories of $^{13}C/^{12}C$ Ratios for Higher Plants. *Plant Physiology* 47:380–384.

Snipp, C. Matthew

2003 Racial Measurement in the American Census: Past Practices and Implications for the Future. *Annual Review of Sociology* 29:563–588.

Somerville, Andrew D., Mikael Fauvelle, and Andrew W. Froehle
2013 Applying New Approaches to Modeling Diet and Status: Isotopic Evidence for Commoner Resiliency and Elite Variability in the Classic Maya Lowlands. *Journal of Archaeological Science* 40(3):1539–1553.
Tieszen, Larry L., and Tim Fagre
1993 Effect of Diet Quality and Composition on the Isotopic Composition of Respiratory CO_2, Bone Collagen, Bioapatite and Soft Tissues. In *Prehistoric Human Bone: Archaeology at the Molecular Level,* edited by J. B. Lambert and G. Grupe, pp. 121–155. Springer-Verlag, Berlin.
Tsuda, Takeyuki, Brenda J. Baker, James F. Eder, Kelly J. Knudson, Jonathan Maupin, Lisa Meirotto, and Rachel E. Scott
2015 Unifying Themes in Studies of Ancient and Contemporary Migrations. In *Migrations and Disruptions: Toward a Unifying Theory of Migration,* edited by T. Tsuda and B. J. Baker, pp. 15–30. University Press of Florida, Gainesville.
Twiss, Katheryn C. (editor)
2007 *The Archaeology of Food and Identity.* Center for Archaeological Investigations, Southern Illinois University Carbondale, Carbondale.
US Census Bureau
1872–1870 Census: Volume 1. The Statistics of the Population. Washington, DC: US Department of Commerce.
van der Merwe, Nikolaas J.
1982 Carbon Isotopes, Photosynthesis, and Archaeology. *American Scientist* 70:596–606.
van Klinken, Gert J.
1999 Bone Collagen Quality Indicators for Paleodietary and Radiocarbon Measurements. *Journal of Archaeological Sciences* 26:687–695.
Voss, Barbara L., J. Ryan Kennedy, Jinhua Selia Tan, and Laura W. Ng
2018 The Archaeology of Home: Qiaoxiang and Nonstate Actors in the Archaeology of the Chinese Diaspora. *American Antiquity* 83(3):1–20.
Williams-Forson, Psyche
2008 More than Just the "Big Piece of Chicken": The Power of Race, Class, and Food in American Consciousness. In *Food and Culture: A Reader,* edited by C. M. Counihan and P. Van Esterik, pp. 342–353. Routledge, London.
White, Christine D., Rebecca Storey, Fred J. Longstaffe, and Michael W. Spence
2004 Immigration, Assimilation, and Status in the Ancient City of Teotihuacan: Stable Isotopic Evidence from Tlajinga 33. *Latin American Antiquity* 15(2): 176–198.
Whitehead, Tony L.
1992 In Search of Soul Food and Meaning: Culture, Food, and Health. In *African Americans in the South: Issues of Race, Class, and Gender,* edited by H. A. Baer and Y. Jones, pp. 94–110. University of Georgia Press, Athens.

14

Paleopathology and Nonspecific Stress Indicators at Point San Jose

COLLEEN F. MILLIGAN, KRISTEN A. BROEHL, KELSIE HART,
VANESSA C. REEVES, KARIN WELLS, AND ERIC J. BARTELINK

Paleopathology is the study of disease from ancient and historical human remains. The term encompasses both a broad understanding of disease processes as they affect skeletal remains and the timeframe or context in which they occur (Grauer 2012). Paleopathological studies are concerned with conditions that lead to osseous and dental changes in archaeological human remains. The importance of such research lies not in what is observed, but how these observations are interpreted. Studies that address the growth and development of individual remains, the presence of trauma, and identifiable disease processes provide evidence of skeletal health and physiological stress during an individual's lifetime. Paleopathology can also provide information on the variation in pathological conditions seen in different contexts. As Ortner stated (2011), the study of paleopathology should contribute to an understanding of how disease shaped social environments, past and present.

The main questions for the paleopathological analysis of the Point San Jose remains concerned what could be said about the origin of the sample and the lives of these individuals interred in the pit. While most paleopathological studies focus on changes at the individual level, the Point San Jose assemblage provides a unique and challenging perspective on paleopathology because the collection consists of mostly disarticulated and commingled skeletal elements. Therefore, the paleopathological focus adopted here concentrates on the osseous and dental changes observed, and less on specific diagnoses of the etiology of those changes.

Paleopathological Methods

During the initial inventory, Point San Jose elements were examined for possible pathological and anomalous changes. Elements with suspected pathological changes were assessed using radiographic or microscopic analysis. Elements with indications of pathology were categorized following recommendations from *Standards for Data Collection from Human Skeletal Remains* (Buikstra and Ubelaker 1994). All statistical calculations were conducted using SPSS Statistics v. 22.

The elements included in this study were those identified as either a specific bone or tooth, or identified by a specific anatomical region. Ribs and vertebrae presented the greatest analytical challenge. During the initial analysis, both ribs and vertebrae were sorted into lots. These lots denoted classes of vertebrae and ribs that were divided by completeness and evidence of pathology. The vertebrae were identified by type (cervical, thoracic, and lumbar), position (1st cervical, 2nd cervical, and 3rd-6th cervical neural arches), and assessed for completeness (neural arches only, bodies only, and complete elements), pathology, and taphonomic conditions. For example, one vertebral lot could be "lumbar neural arches" while another could be "lumbar pathology." Likewise, the ribs were identified by number (for the first rib only), side (left, right, and unsided), and assessed for completeness (% complete), pathology, and taphonomic conditions. For example, one rib lot could be "right ribs 25–75% complete" while another could be "right Rib 1, pathology."

To minimize the possibility of recording multiple observations for a single fragmented element, the lots included in the paleopathology examination were elements in good or excellent condition and nearly complete. The observations of pathology noted in the lots were compared to the total number of bones in the collection complete enough for pathological assessment. Fifty percent of the total collection was included in the examination of pathological conditions (2,231 fragments out of 4,450 fragments).

The second analysis tested the reliability of the initial paleopathology inventory. Specifically, all pathological vertebrae and rib lots were reexamined for consistency using Cohen's kappa (k), a measure of inter-rater agreement for categorical data. The results indicated no significant differences. In addition, 130 vertebral fragments and 80 rib fragments from nonpathological lots were randomly sampled for reexamination. A total of 13 vertebral fragments and 8 rib fragments showed previously unrecorded signs of pathology. However, this result also showed no statistically significant differences between the primary and secondary analysis. These conditions ranged from mild to moderate expressions of lipping, porosity, enthesophyte formation, or periosteal reactions. No additional instances of pathognomic diseases or trauma were observed.

This second analysis showed that the reliability of the initial inventory was high given that the reassessment did not produce significant differences in prevalence rates. The degree of fragmentation of vertebrae and ribs was substantial, causing an increased number of fragments relative to the number expected based on anatomical representation (Cox et al., this volume). This increased fragmentation led to an increased possibility of overlooking pathological conditions due to both size of the fragments and loss of identifiable morphology in many cases. It highlights the utility of conducting multiple analyses of material in paleopathological study, especially when high levels of fragmentation are seen.

The overall methodology used for the paleopathological observations, regardless of element type, recognized that each element might show multiple pathological conditions. Each observation noted the presence or absence of three general expressions of pathology: shape or size change, bone loss or formation, and possible disease or trauma. Shape changes included abnormal curvature of postcranial elements and premature suture closure of cranial elements. Size changes consisted of atrophied postcranial elements and abnormal calvarium size. Bone loss could be focal or diffuse, while bone formation consisted of woven and sclerotic patterns. Bone formation that occurred in muscle tissue (e.g., myositis ossificans) or connective tissue was noted. Ortner (2011) states that these four abnormalities (shape, size, bone formation, and bone loss) are the ways that the skeleton shows change, and they are also abnormalities that may overlap.

Possible disease or traumatic conditions were further classified as trauma, infection, metabolic disease, degenerative disease, and dental disease. Because trauma in the Point San Jose materials was exclusively bone fractures, this category was assigned when a fracture was observed and confirmed by radiographic analysis. Infection was identified when an element was greater than 75 percent complete and bony reaction occurred without traumatic injury. Metabolic disease was indicated when porosity was observed in the superior roof of the eye orbits or cranial vault (cribra orbitalia and porotic hyperostosis, respectively). Joint disease was identified when articular surfaces showed evidence of bone loss or formation indicating degeneration (e.g., eburnation, osteophytic lipping, and/or porosity on the joint surface).

Analysis of dental disease included dental caries, dental wear, calculus deposition, and linear enamel hypoplasia. For this study, the prevalence of dental caries was assessed in all teeth following the protocol from *Standards for Data Collection* (Buikstra and Ubelaker 1994). Dental wear scores were recorded using the Smith (1984) and Scott (1972) systems described in *Standards* (Ubelaker and Buikstra 1994). Dental calculus was also scored following *Standards.*

Linear enamel hypoplasia defects (LEH) were observed macroscopically. To estimate the age when the defect occurred, the distance of LEH from the cemento-enamel junction was measured to the nearest .02 mm following *Standards* (Buikstra and Ubelaker 1994). The age of defect formation was estimated using regression formulae from Goodman and Song (1999), which were derived from mean crown heights in Swärdstedt (1966) and dental development sequences in Massler et al. (1941). LEH age of formation was calculated for adult canines only. Due to the commingled nature of the dental remains and the number of loose teeth, each tooth was scored individually and tabulations were made by tooth type.

The analysis also compared the prevalence and type of pathology seen with other samples. Two samples were chosen for comparison with the osseous paleopathological results from the Point San Jose sample: the Medical College of Georgia (1835–1913, Harrington 1997) and the Holden Church (1800–1862, Hodges et al. 2017). As with Point San Jose, both of the other samples represent rare incidents of excavated medical or anatomical waste. All three samples are urban sites and roughly contemporaneous. All three samples are thought to have a connection to medical training either through their physical location or through the suspected practitioner. All samples are also represented by individual, commingled elements rather than by complete individuals.

Because bioarchaeological studies often report dental results by individual tooth prevalence rates, the commingled and disarticulated nature of the Point San Jose sample was not as significant a limitation as it was for the osseous analysis. The dental comparisons employed more temporally and geographically diverse groups of historic military and civilian samples: a British historic sample (eighteenth–nineteenth century, Wells 1968); Snake Hill (1812–1814), Indian Wars (1870–1899), and Civil War (1861–1865, Sledzik and Moore-Jansen 1991); St. Peter's (1720–1810, Owsley et al. 1987); and Fort Laurens (1778, Sciulli and Gramly 1989). The paleopathological results are divided by osseous and dental analyses, and the results of each section are discussed in turn.

Paleopathology: Osseous Results

Out of 2,231 elements examined, 447 (20.0%) showed evidence of paleopathological conditions. When divided by anatomical region, pathology was observed in 19 elements of the skull (4.3%), 192 of the vertebrae (43.0%), 39 of the rib cages (8.7%), 18 of pelves (4.0%), 40 of the arms (8.9%), 39 of the legs (8.7%), 41 of the wrists/hands (9.2%), and 59 of the ankles/feet (13.2%). The vertebrae alone accounted for 43.0 percent of all elements with pathology. The wrists/hands and ankles/feet combined represented 22.4 percent of all elements with pathological conditions.

Shape and Size Change

Of the 447 total elements with evidence of pathology, 22 (4.9%) exhibited a change in shape, but no elements displayed changes in size. All 22 elements with shape changes were in the postcranial skeleton. Twelve of the observed shape changes (54.5%) were associated with trauma to the element.

Bone Loss and Formation

Pathological changes involved bone loss, bone formation, or both. A total of 51 (11.4%) elements of 447 with pathology exhibited bone loss and 361 (80.8%) of 447 exhibited bone formation. Nine elements showed evidence of both loss and formation. Bone formation was by far the most common pathological change and often indicated degenerative disease, particularly in the vertebrae.

Trauma

As noted earlier, traumatic observations consisted entirely of fractures and all were confirmed radiographically. A total of 24 (5.4%) of 447 pathological elements exhibited either complete or incomplete healed fractures. Seven of the fractures were associated with degenerative joint changes (e.g., osteoarthritis secondary to trauma). All fractures showed evidence of healing, and therefore were also scored for bone formation. Under normal circumstances, bones begin fracture repair with a soft callus. A secondary, hard callus comprising new bone material subsequently begins to replace the initial, stabilizing soft callus. Over time, the new bone created as part of the healing process is remodeled and reorganized. All observed Point San Jose fractures were in the remodeling stages of healing. Figure 14.1 shows four exemplars of fractures seen in the Point San Jose sample, representing transverse, oblique, and displaced shear fracture patterns.

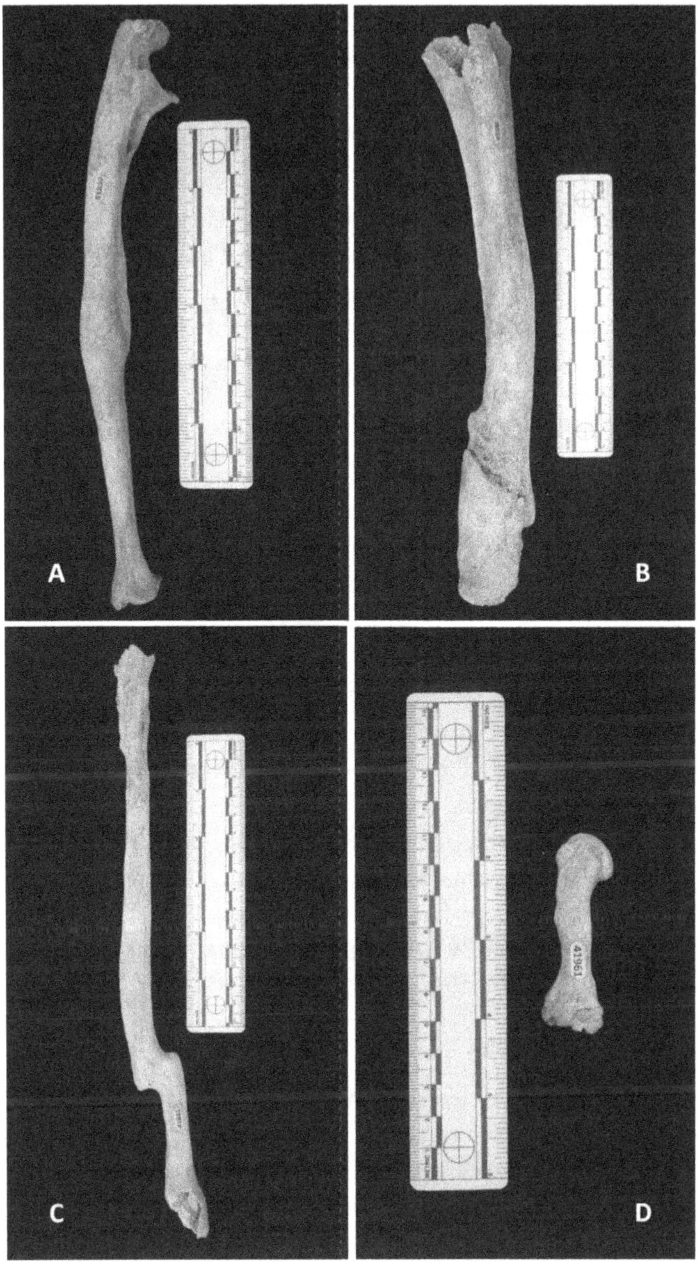

Figure 14.1. Point San Jose traumatic alterations. (*A*) Left ulna (Catalog 41859). Transverse mid-shaft fracture showing healing. (*B*) Right tibia, anterior view (Catalog 41880). Shear fracture of the distal shaft with poor alignment and healing. (*C*) Right fibula (Catalog 41881). Comminuted shear fracture of the proximal and distal shaft with healing. (*D*) Left metacarpal 2 (Catalog 41961). Oblique fracture with traumatic joint disease at the proximal end showing healing. Courtesy of National Park Service and Chico State.

Infection

Of the 447 elements with pathology, 6 (1.3%) showed evidence of infection. These conditions consisted primarily of periosteal reactions. Periosteal reactions result from an inflammation of the periosteum that may be caused by disease, trauma, or even developmental factors (Ortner 2003). The Point San Jose elements that displayed periosteal reactions but lacked evidence of trauma were considered evidence of nonspecific infection. No differential diagnoses of the periosteal reactions were made because the focus of the investigation was on single elements instead of individuals.

Metabolic Disease

Metabolic disease included skeletal changes usually associated with vitamin deficiencies or acquired anemias. Two (0.45%) elements of the 447 indicated metabolic disease. Both examples consisted of parietal bones that showed porosity of the cranial vault, consistent with healed porotic hyperostosis.

Degenerative Disease

A total of 303 (67.8%) elements of 447 with pathology showed evidence of degenerative changes. Of these, 297 elements demonstrated bone formation as lipping or osteophytic growth of joint surfaces, and 16 evidenced bone loss associated with porosity, depressions, or eburnation of joint surfaces. As noted in the trauma section, 13 of the elements with joint changes displayed trauma. Degenerative joint disease in the vertebrae was particularly notable, constituting 39.2 percent of all affected joints in the Point San Jose collection.

Radiographic Assessment

The authors examined 30 elements using radiographs during the secondary analysis. Elements selected to be radiographed showed trauma or ambiguous osseous changes. For example, Figure 14.2 shows a radiographic exemplar of traumatic injuries. The subsequent analysis resulted in three possible conclusions: 1) the pathological changes were associated with trauma, 2) ambiguous pathological changes were most likely traumatic in origin, or 3) preliminary pathological observations were revised. Eight elements of the 30 radiographed had been originally identified as having traumatic injuries, but with additional information were classified as non-traumatic. These eight showed either non-traumatic shape changes or infection as

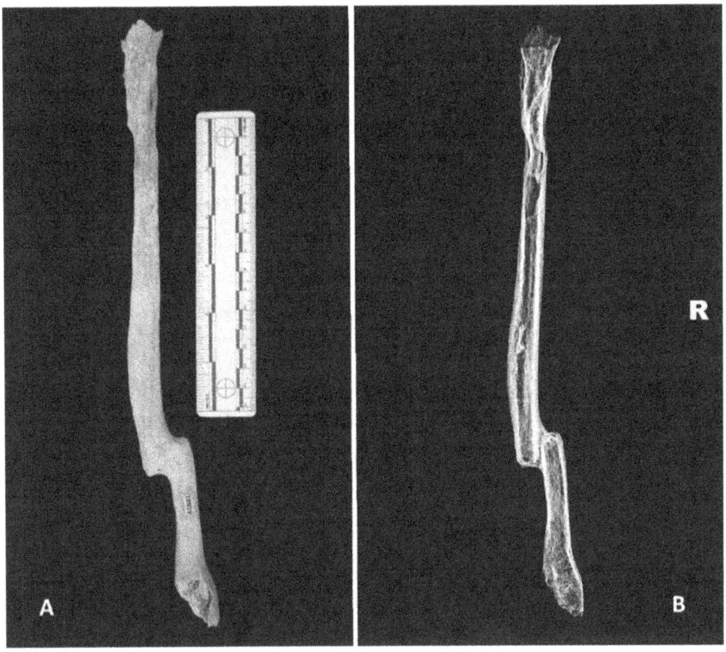

Figure 14.2. Point San Jose healed fracture. Displaced, transverse fracture of a right fibula (Catalog 41881) (*A*) photograph and (*B*) radiograph. Courtesy of National Park Service and Chico State.

more likely causes of a bony reaction. Combining a traditional macroscopic examination of the elements with a radiographic examination results in paleopathological conclusions with multiple lines of evidence. In this case, it resolved ambiguity from the visual assessment of pathology alone.

Comparative Osseous Results

Two samples were chosen for comparison with the osseous paleopathological results from the Point San Jose sample: the Medical College of Georgia sample and the Holden Church sample. The Medical College of Georgia sample was discovered in 1989 during construction in the basement of the Old Medical College Building. The building housed the college's anatomy and pathology laboratories between 1835 and 1913 (Blakely 1997). Harrington (1997) reported that 16.8 percent (1,643/9,808) of the elements found in the basement of the Medical College of Georgia showed evidence of pathological conditions. The sample was examined for trauma, infectious disease (specific and nonspecific), metabolic disease, growth disorders, osteoporosis, degenerative joint disease, and oral pathology. The most frequent pathological conditions were osteoporosis and degenerative joint disease (Harrington 1997).

Similar to the Medical College of Georgia sample, the Holden Chapel sample was discovered in 1999 during construction work in a basement. Human remains and other artifacts were found in a well located in the basement. The building was used by Harvard's Medical School for anatomy lectures between 1800 and 1862 (Hodge et al. 2017). Hodge et al. (2017) reported that 5.3 percent (48/907) of the elements found at Holden Chapel showed evidence of pathological conditions with osteoarthritis and trauma being the most frequent.

When looking at the osseous rates of pathology alone, 20 percent (447/2,231) of the Point San Jose sample showed evidence of pathology. Data extrapolated on the osseous remains of the Holden Chapel sample indicate 5.2 percent (46/877) exhibited evidence of pathology, and that 14.7 percent (1,445/9,808) of the Medical College of Georgia sample had evidence of pathology. A significant difference was found between samples for prevalence of pathology (χ^2= 84.672, $p < 0.001$). In a pairwise assessment of risk for each sample, the odds ratio for Point San Jose and Holden Chapel showed that Point San Jose was 3.8 times more likely to demonstrate pathology than Holden Chapel. Likewise, Point San Jose was 1.4 times more likely to show pathology than Medical College of Georgia. Given that several pathological elements could have originated from the same individual, these prevalence rates are not the prevalence of the pathological conditions in the sample or even population; instead they show the frequency of inclusion of pathological-to-non-pathological elements in the samples.

The data for all three samples were largely consistent. The types of pathology were grouped into four broader categories: trauma, infection, metabolic disease, and degenerative disease. While these categories do not encompass all of the variation in pathology observed, they do include 50 percent or greater of the pathological conditions for each sample. Again looking at the compiled osseous results, prevalence for each category of pathology was calculated for each sample. Figure 14.3 shows the prevalence rates for each category and each sample.

As noted earlier, degenerative disease accounts for the largest percentage of pathology in each of the samples. Both Point San Jose and Holden Chapel show trauma as the next highest category while the Medical College of Georgia sample demonstrated infection as the second highest in frequency. For all samples, metabolic disease displayed the lowest percentage of pathology. Chi-square and odds ratios were calculated for each cat-

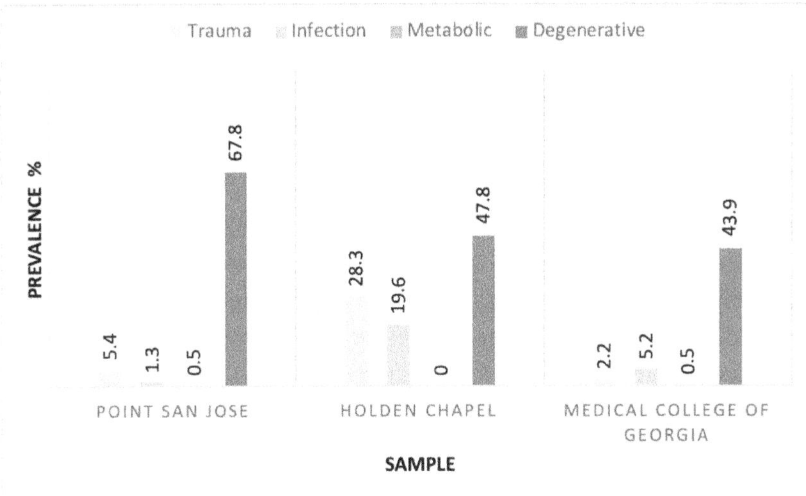

Figure 14.3. Comparison of pathology categories among the Point San Jose, Holden Chapel (Hodge et al. 2017), and Medical College of Georgia (Harrington 1997) samples.

egory of pathology except metabolic disease. For each of the three samples, metabolic disease was too small of a sample size to provide meaningful comparisons. For all other categories of pathology, significant relationships were found between the samples: trauma (χ^2= 93.619, $p < 0.001$), infection (χ^2= 35.105, $p < 0.001$), and degenerative disease (χ^2= 60.283, $p < 0.001$).

Both Point San Jose (2.5 times) and Holden Chapel (17.4 times) showed a greater risk for trauma than Medical College of Georgia. This result was not surprising considering that both Point San Jose and Holden Chapel had higher frequencies of trauma than the Medical College of Georgia. Holden Chapel displayed 4.4 times greater risk for infection than the Medical College of Georgia sample as well. Point San Jose showed two times greater risk for degenerative disease than the Holden Chapel sample, and 2.4 times greater risk for the Medical College of Georgia sample. While all samples exhibited the highest frequency for degenerative disease, Point San Jose showed a high degree of fragmentation of the axial skeleton, which contributed to a higher number of axial fragments than either of the comparative samples. The axial skeleton also displayed a high frequency of degenerative disease. It would be likely, given the degree of fragmentation, that at least some of the axial fragments are from the same individual. That association may explain why Point San Jose shows a higher risk estimate than both Holden Chapel and Medical College of Georgia in this category. All other risk estimates between samples were negligible.

Paleopathology: Dental Results

Pathological conditions of the jaws and teeth provide insight into the diet and oral health of human societies. The interaction between teeth and food provide the basis for studying the impact of diet on dental health, including dental caries, occlusal wear, and alveolar abscesses (Hillson 1996, 2001). Other observations of dental disease and modifications included dental caries, dental wear, dental calculus, and linear enamel hypoplasias. Dental pathological conditions, unlike bony changes, are often presented by tooth type. Consequently, the single tooth observations of the Point San Jose dental material were comparable to many other studies.

Dental Caries and Calculus

For this study, all carious lesions were grouped together and prevalence was calculated by tooth type. In total, 31 incisors, 23 canines, 52 premolars, and 59 molars were observed for dental caries (n = 165). The overall dental caries prevalence is 11.5 percent, with the molars (23.7%) showing the highest rate of carious lesions, followed by the canines (9.1%), incisors (6.5%), and premolars (1.9%). The prevalence of dental caries by tooth type is statistically significant ($\chi^2 = 13.897$, $p = 0.003$). Dental calculus was common, observed on 66.1 percent (109/165) of teeth. However, no relationship was found between caries and calculus prevalence ($\chi^2 = 0.005$, $p = 0.943$).

Linear Enamel Hypoplasias

In total, 19.4 percent (32/165) of Point San Jose teeth showed at least one LEH defect. Among teeth having LEHs, the minimum number of defects was one and the maximum number was five. The highest prevalence of teeth with LEH defects occurred among incisors (15/31, 48.4%) and canines (11/23, 47.8%), followed by premolars (4/52, 7.7%), and molars (2/59, 3.4%), a statistically significant difference ($\chi^2 = 42.784$, $p < 0.001$).

The average age of canine LEH formation was 4.3 ± 0.7 years, with the youngest defect occurring at 2.94 years and the oldest defect occurring at 5.4 years. Some of the mandibular canine defects had estimated ages of formation outside the tooth formation window of 4.5 years; however, this apparent discrepancy may have resulted from slower dental maturation rates in these individuals. The Point San Jose LEH data suggested that repeated episodes of nutritional or disease-related stress was a common occurrence during early childhood for at least some of the individuals in the sample.

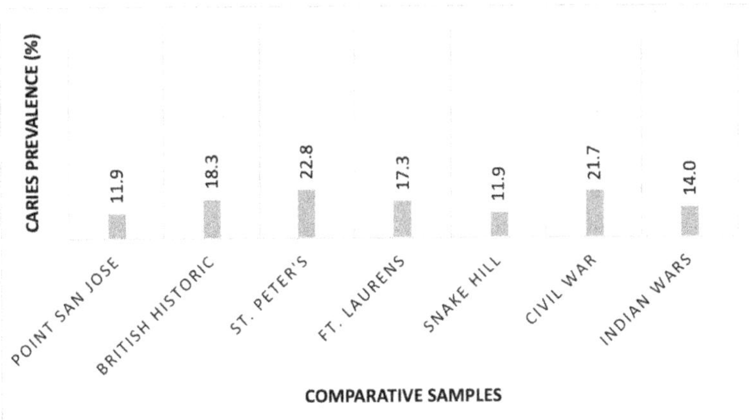

Figure 14.4. Comparison of dental caries prevalence between the Point San Jose dental sample and six other historical period samples. Sites: Point San Jose (1870s); British Historic (eighteenth–nineteenth century): Wells (1968); Snake Hill (1812–1814), Indian Wars (1870–1899), and Civil War (1861–1865): Sledzik and Moore-Jansen (1991); St. Peter's (1720–1810): Owsley et al. (1987); Fort Laurens (1778): Sciulli and Gramly (1989).

Comparative Dental Caries Results

The dental caries rate among the Point San Jose individuals, especially for molars, suggests a fairly cariogenic diet (high in sugar and other carbohydrates) and poor overall oral health. These values are comparable to archaeological agricultural populations (Larsen et al. 1991). Figure 14.4 shows comparative data from a number of historic samples (Owsley et al. 1987; Sciulli and Gramly 1989; Sledzik and Moore-Jansen 1991; Wells 1968). Comparison of the Point San Jose with other historical period samples indicates a similar caries prevalence as the Snake Hill Site (1812–1814) and Indian Wars sample (1870–1899), but a much lower rate than a British historical sample and comparative historical samples from St. Peter's (1720–1810), Fort Laurens (1778), and the Civil War (1861–1865). The comparative samples combine military and civilian samples and represent oral health patterns throughout the historical period.

Discussion

The question of "what does any of it mean" is at the core of any paleopathological analysis. When samples consist of distinguishable individuals, this question is still difficult to address. As emphasized by Wood et al. (1992), caution should be taken in any paleopathological study. No sample is fully representative of the population from which it comes nor does it present an accurate reflection of the stress and disease burden that individuals experienced based on skeletal indicators alone (Wood et al. 1992). This limitation is even more evident when pathology is assessed by individual skeletal elements. In collections like Point San Jose, researchers are blind to whether or not multiple elements showing pathology come from the same individual unless the elements can be articulated. The inaccessibility of pathological patterns *within* individuals further reduces the ability to calculate prevalence or risk from the sample.

The first of the main research questions for this analysis concerned the possible origin of the sample. Overall, the Point San Jose results indicate about 20 percent of elements showed pathological changes. The highest prevalence was for bone formation associated with joint disease. Other pathological conditions, such as infection, shape changes, and trauma, were present in the assemblage but at relatively low levels. This result reinforces the idea that there was no selection for noteworthy pathological conditions in the recovered remains. It is possible that noteworthy pathological specimens could have been retained from the Point San Jose deposit and used for demonstration or curation elsewhere, but the remains discarded in the Point San Jose pit are not representative of a curated collection.

For example, two eighteenth and nineteenth century British samples of pathological teaching collections show a higher level of significant pathology. The first sample is a small archaeological assemblage from Surgeon's Square that included 47 human bones with cut marks. This sample was found on the grounds of an eighteenth-to-nineteenth century anatomy school in Edinburgh, Scotland, and represented a pathological teaching collection. There were at least five individuals and 34 percent of elements showed severe pathology, including osteomyelitis, Paget's Disease, tuberculosis, and trauma. A pin in the tubercular vertebrae supports the use of the remains as a demonstrative pathological specimen (Henderson et al. 1996).

Another small eighteenth-to-nineteenth century pathological collection was discovered near a hospital in Nottingham, United Kingdom. At least five individuals were also present and 25 percent of elements showed severe pathological conditions, such as rickets, osteomyelitis, achondroplasia, and septic arthritis. The presence of severe pathological conditions, incised and saw cut marks, and a hinge on one element also support the probability that this series was a pathological teaching collection (Chapman 1997).

Both of the Surgeon's Square and the Nottingham collections show higher rates of pathology and types of pathological conditions that would be of clinical interest. Evidence that these specimens are representative of a curated or teaching collection is also supported by the recovery context of these two samples. The paleopathology seen within Point San Jose sample is more consistent with the collection or medical waste of dissected cadavers or surgical patients. There could still be some elements in the collection that functioned as pathological teaching specimens, because multiple types of medical discard could be present in the pit, but the overall paleopathological analysis suggests that the majority of the collection was not used in this way.

The second primary question behind the paleopathological analysis takes a much more holistic approach. It concerned what can be inferred about the lives of the individuals that comprise the sample. This chapter is one of many in this volume that adds information to the overall question of identity. As noted several times, the fact that the collection is commingled and disarticulated greatly limits how much information can be inferred from the skeletal remains. The samples can only look at individual skeletal elements instead of the individuals themselves.

The addition of demographic information usually enhances the inferences about individuals in archaeological samples. Point San Jose, as other probable medical waste samples, presented difficulty in assessing demographic information for a large portion of the individuals. However, understanding the demography of the sample, even limited, is an important starting point for understanding the meaning of the pathological conditions observed in the sample. The demographic information for Point San Jose (Sgheiza and Willey, this volume) shows elements that range in age from fetal to adult but with more adults represented overall. A similar age profile was also seen with the comparative samples: Holden Church and Medical College of Georgia (Harrington 1997; Hodge et al. 2017). Likewise, all three samples have both males and females with a bias toward males (Harrington 1997; Hodge et al. 2017; Sgheiza and Willey, this volume). The underlying similarity between the samples may be that they would seem to represent marginalized groups in nineteenth-century America. As discussed by Hodge et al. (2017) regarding the Holden Church sample, the overall assessment of the pathology suggests that the individuals recovered from Point San Jose were opportunistically collected for either dissection or the remains were part of medical waste associated with the fort hospital.

Conclusions

Overall, the Point San Jose elements showcase a variety of paleopathological conditions, both osseous and dental. However, the primary pathological condition was degenerative disease. When compared to similar samples such as the Medical College of Georgia and Holden Chapel, Point San Jose showed an increased risk for degenerative disease overall. All three samples had degenerative disease as the most common paleopathological observation. For primarily adult skeletal samples, observations of degenerative disease are common and thus not surprising as the leading pathological condition. Likewise, the trauma and infection in the Point San Jose collection are also common observations. Ortner (2003) states that degenerative conditions, trauma, and infection are the three most common occurrences of pathology in archaeological contexts.

The methodology used in the Point San Jose dental assessment was comparable to other studies of nineteenth-century skeletal series because dental analysis is most often reported by tooth type instead of the dental arcade as a whole. As a result, the conclusions drawn are better comparisons of the overall health than the osseous changes. LEHs occur in 32 of 165 teeth (19.4%). Most of these teeth display multiple defects, suggesting repeated early-childhood stress among at least some of the Point San Jose individuals. In addition, 11.5 percent of the teeth showed carious lesions. However, evidence for dental disease was lower than other contemporary historic samples suggesting that oral health was typical of the historical period when sugar and other carbohydrates were more commonly available.

The most significant conclusion drawn is that the Point San Jose paleopathology represents expected observations for a historic sample. The paleopathological analysis of the remains associated with the Point San Jose sample does not necessarily show why the individuals were chosen from a pathological standpoint. However, their very inclusion in the sample, outside of a normal burial setting and with low levels of overall significant pathology, may suggest their context as possibly marginalized members of society at the time. The question of identity of the Point San Jose individuals is expanded in subsequent chapters, but the paleopathological analysis adds important contextual information to the significance of this assemblage.

References Cited

Blakely, Robert L.
1997 A Clandestine Past: Discovery at the Medical College of Georgia and Theoretical Foundations. In *Bones in the Basement: Postmortem Racism in Nineteenth-Century Medical Training,* edited by Robert L Blakely and Judith M. Harrington, pp. 3–27. Smithsonian Institution Press, Washington, DC.

Buikstra, Jane E., and Douglas H. Ubelaker
1994 *Standards for Data Collection from Human Skeletal Remains.* Arkansas Archaeological Report Series, 45. Arkansas Archaeological Survey, Fayetteville.

Chapman, S. J.
1997 The Findings of a Possible Reference Collection in the Grounds of a Victorian General Hospital, Nottingham, UK. *Journal of Paleopathology* 9:37–46.

Goodman, Alan H., and R. J. Song
1999 Sources of Variation in Estimated Ages at Formation of Linear Enamel Hypoplasias. In *Human Growth in the Past: Studies from Bones and Teeth,* edited by Robert D. Hoppa and Charles M. Fitzgerald, pp. 210–240. Cambridge University Press, Cambridge, United Kingdom.

Grauer, Anne L.
2012 Introduction: The Scope of Paleopathology. In *A Companion to Paleopathology,* edited by Anne L. Grauer, pp. 1–14. Blackwell Publishing, West Sussex, United Kingdom.

Harrington, Judith M.
1997 Death and Disease: The Paleopathology of the Medical College of Georgia Cadaver Sample. In *Bones in the Basement: Postmortem Racism in Nineteenth-Century Medical Training,* edited by Robert L. Blakely and Judith M. Harrington, pp. 261–302. Smithsonian Institution Press, Washington, DC.

Henderson, David, Mark Collard, and Daniel A. Johnston.
1996 Archaeological Evidence for 18th-century Medical Practice in the Old Town of Edinburgh: Excavations at the 13 Infirmary Street and Surgeons' Square. *Proceedings of the Society of Antiquaries of Scotland* 126:929–941.

Hillson, Simon
1996 *Dental Anthropology.* Cambridge University Press, New York.
2001 Recording Dental Caries in Archaeological Human Remains. *International Journal of Osteoarchaeology* 11:249–289.

Hodge, Christina J., Jane L. Rousseau, and Michèle E. Morgan
2017 Teachings of the Dead: The Archaeology of Anatomized Remains from Holden Chapel, Harvard University. In *The Bioarchaeology of Dissection and Autopsy in the United States,* edited by Kenneth C. Nystrom, pp. 115–139. Springer International, Cham, Switzerland.

Larsen, Clark Spencer, Rebecca Shavit, and Mark C. Griffin
1991 Dental Caries Evidence for Dietary Change: An Archaeological Context. In *Advances in Dental Anthropology,* edited by Mark A. Kelley and Clark Spencer Larsen, pp. 179–202. Wiley-Liss, New York.

Massler, Maury, Isaac Schour, and H. G. Poncher

1941 Developmental Pattern of the Child as Reflected in the Calcification Pattern of the Teeth. *American Journal of Disease of Children* 62:33–67.

Ortner, Donald J.

2003 Identification of Pathological Conditions in Human Skeletal Remains. Academic Press, New York.

2011 Human Skeletal Paleopathology. *International Journal of Paleopathology* 1:4–11.

Owsley, Douglas W., Charles E. Orser, Jr., Robert W. Mann, Peer H. Moore-Jansen, and Robert L. Montgomery

1987 Demography and Pathology of an Urban Slave Population from New Orleans. *American Journal of Physical Anthropology* 74:185–197.

Scott, Eugenie C.

1972 Dental Wear Scoring Technique. *American Journal of Physical Anthropology* 51:213–218.

Sciulli, Paul W., and Richard M. Gramly

1989 Analysis of the Ft. Laurens, Ohio, Skeletal Sample. *American Journal of Physical Anthropology* 80:11–24.

Sledzik, Paul S., and Peer H. Moore-Jansen

1991 Dental Disease in Nineteenth Century Military Skeletal Samples. In *Advances in Dental Anthropology,* edited by Mark A. Kelley and Clark Spencer Larsen, pp. 215–224. Wiley-Liss, New York.

Smith, B. Holly

1984 Patterns of Molar Wear in Hunter-Gatherers and Agriculturalists. *American Journal of Physical Anthropology* 63:39–56.

Swärdstedt, Torsten

1966 Odontological Aspects of a Medieval Population in the Province of Jämtland, Mid-Sweden. PhD dissertation, Anthropology Department, University of Stockholm.

Wells, Calvin

1968 Dental Pathology from a Norwich, Norfolk, Burial Ground. *Journal of the History of Medicine and Allied Sciences* 23(4):372–379.

Wood, James W., George R. Milner, and Henry C. Harpending

1992 The Osteological Paradox: Problems of Inferring Prehistoric Health from Skeletal Samples. *Current Anthropology* 33:343–370.

IV

Conclusions

15

Bodies Apart

Structural Violence and Dis/Embodiment at Point San Jose

SARAH A. HALL

Nineteenth-century skeletal samples from the US routinely reflect marginalized and vulnerable populations, some of which were also subject to dissection (see Nystrom 2014). At the time, anatomization and dissection were highly stigmatized, even considered a form of desecration. Using historical and osteological data, this work builds upon previous scholarship to discuss the marginalized social identities of the individuals in the skeletal assemblage at Point San Jose (1860–1900) in San Francisco, California. Here, I explore the ways structural violence predisposed certain segments of the population to dissection, to inclusion within the kinds of burials likely to be forgotten (much less rediscovered), and to destructive methods of excavation and analysis. The ensuing discussion of embodiment and postmortem agency is applied not only to historical postmortem examinations but also to the excavation and analytical methods used for this assemblage.

Previous studies have described historical evidence of both anatomical dissection and anatomical collection in terms of embodiment theory, structural violence, and structural inequality (de la Cova 2019; Nystrom 2011, 2014, 2017). Evidence from skeletal analysis and sociohistorical context is used here to demonstrate how the remains interred at Point San Jose reflect the structural violence experienced during life and that continued in death. Furthermore, this study uses evidence of postmortem examination, specifically dissection, to illustrate disembodiment and postmortem agency in this assemblage. This latter theme of postmortem agency persists throughout the death histories of these individuals, as dead bodies continue to impact individuals today, particularly bioarchaeologists. Implications for ethical bioarchaeological practice among marginalized groups are considered in light of postmortem agency and structural inequality.

Structural Violence Theory

The term "structural violence" was coined by Johan Galtung (1969), a sociologist and mathematician. In this foundational work, Galtung (1969) expanded the meaning of violence beyond mere physical harm, beyond the work of actors, and beyond intention. Under this definition, violence is done when avoidable obstacles prevent individuals from attaining their potential (Galtung 1969, 1975). Structural violence occurs when asymmetrical power structures harm or disadvantage people. In this way, indirect structural violence is distinguished from direct physical violence.

Structural violence pertains to unequal distributions of power and resources and does not require an actor—it is the structure that enacts violence, not a specific person. Various social structures may be responsible for this violence, including economic, cultural, political, or legal institutions (Rylko-Bauer and Farmer 2016). Cultural violence refers to the aspects of culture that legitimize harm, making direct and indirect forms of violence (including structural violence) "look and even feel right" (Galtung 1990:291). Forms of structural and cultural violence are self-perpetuating; they reinforce existing dynamics of power, which further necessitate these forms of violence to maintain legitimacy.

Anthropological studies of structural violence are predominately concentrated in the sub-disciplines of medical and cultural anthropology in the work of scholars like Scheper-Hughes (1992), Farmer (1996, 2003, 2004), and Bourgois (2009). Whereas Galtung (1969, 1975) and his contemporaries were preoccupied with quantitatively evaluating structural violence (e.g., Köhler and Alcock 1976), more recent work by Farmer (1996, 2003, 2004) distinguished structural violence from other forms of violence by focusing on which social institutions cause human suffering and how social inequality is embodied. Farmer framed anthropological research in terms of structural violence and social medicine, emphasizing the local, national, and international political and economic structures that disproportionately affect the health of the most vulnerable groups within a population, substantiated elsewhere by other researchers (Bourgois 2009; Quesada et al. 2011; Singer and Clair 2003). Further, Scheper-Hughes and Bourgois (2004) suggested a continuum of violence to explain how daily injustices (structural violence) predispose groups to more direct forms of violence.

Structural violence is also a useful framework in the subfield of biological anthropology. When a biocultural approach is employed to understand patterns of disease and violence, researchers have effectively examined the embodied biological effects of structural processes (Blakey 1998; Goodman 2013; Goodman and Leatherman 1998; Martin et al. 2008). Additionally, many bioarchaeologists have explicitly studied embodied structural violence and institutional inequality in skeletal assemblages and skeletal collections (e.g., Bright 2020; de la Cova 2019; Harrod et al. 2012; Klaus 2012; Lans 2018; Nystrom 2011, 2014, 2017; Tremblay and Reedy 2020; Watkins 2018).

Structural Inequality and Social Marginalization

Structural inequality and social marginalization are aspects of structural violence and are particularly relevant to the discussion of nineteenth-century postmortem examination. Any structure can produce violence against a group of people, and in fact, many different structures oppressed numerous groups of people in different ways to put them at risk of dissection in the late nineteenth century. Here, I explore the outcomes of that structural violence—inequality and social marginalization—and their relationship to partible practices.

Postmortem examinations include both autopsy and dissection, but the two had starkly different connotations in the nineteenth century. Autopsies were rare, and in some ways, it was considered an honor to be important enough to merit an autopsy (Crossland 2009; Sappol 2002). Dissection, on the other hand, was considered a violation and a kind of punishment (Garment et al. 2007; Humphrey 1973; Sappol 2002; see Milligan, this volume). In the nineteenth century, "Death had a capital D and served a moral purpose" (Sappol 2017:vii), and death's moral valence meant that it was attended by considerable ritual, more or less elaborate depending upon a person's financial means and social status. Consequently, what happened to the body after a person died was a point of considerable concern and anxiety, especially during the rise of "resurrectionists," more commonly known as grave robbers.

Due to these cultural values around death and the body, there were very few willing anatomical donors available for study in the nineteenth century. Medical students and surgeons had few legitimate means of obtaining cadavers, so they acquired bodies however they could, typically targeting the most vulnerable members of the population. Bodies of executed criminals were (legally) used for dissection, although demand outpaced executions (Richardson 2000; Sappol 2002). Grave robbing spread and preferentially targeted marginalized people from communities with limited ability to seek redress, mostly those buried at Black burial grounds or "pauper" cemeteries (public cemeteries where the deceased were those who could not afford a burial) (Humphrey 1973). Despite the grisly reputation grave robbing brought to dissection, proponents of Bentham's utilitarian philosophy argued that the dead were a resource to the living (Southwood Smith 1824). Indeed, Jeremy Bentham himself donated his own body for dissection in an effort to encourage the practice (Richardson and Hurwitz 1987), initiating a debate about the rights of the dead and researchers' ethical responsibilities that continues today (Garment et al. 2007; Nelkin and Andrews 1998).

Bentham and Southwood Smith helped establish the foundation for the first Anatomy Acts in Britain, which allowed licensed medical professionals legal access to unclaimed bodies from public institutions such as jails, prisons, almshouses, hospitals, and asylums—bodies of so-called paupers who would otherwise be buried with public funds (Garment et al. 2007; Richardson 2000; Richardson and Hurwitz 1987). By the late nineteenth century, most states in the US had passed similar laws. The California Anatomy Act of 1870 allowed physicians access to unclaimed bodies from public institutions. Further, it contained the important (and common) proviso that all bodies used for the anatomical sciences were to be given a proper burial, paid for by the anatomist or surgeon to whom they were given (Cal. Gen. Laws ch. 304, §3, 1870). This was law during the period when the Point San Jose pit was made.

The individuals most impacted by these laws and practices would have been poor and otherwise marginalized groups, including Black and immigrant populations. Socially and economically marginalized individuals are already disproportionately vulnerable to negative health outcomes (Marmot et al. 2008; Singer 2009; Singer et al. 2017), which would have placed these groups at increased risk of reliance on public institutions, and thus increased risk of dissection. Postmortem examination of Black bodies was common and is well-documented (Berry 2017; Humphrey 1973; Savitt 1982), in large part because Black communities had little social protection against exploitation (Davidson 2007; Humphrey 1973). Foreign immigrants were also disproportionally represented at public institutions in the US and often experienced high rates of trauma and pathology (Humphrey 1973).

Marginalized peoples of the nineteenth century and their oppressors were keenly aware of the psychological and spiritual harm done by the threat of dissection. Dissection of bodies from public institutions was considered a kind of punishment for indigence, and even referred to in anatomy laws as a deterrent against poverty (see discussion in Sappol 2002). As a complaint to the Board of Guardians of the Philadelphia almshouse in 1845 noted, "That it occasions dread and anxiety in the minds of some of the inmates of this House, is a well [-] known fact . . . burial here, during the lecture season, is a mockery, and to be buried elsewhere is sometimes asked as the last and greatest favor" (Lawrence 1905:161).

After Life: Dis/Embodiment and Postmortem Agency

The concepts of embodiment, disembodiment, and postmortem agency bridge the boundary between the realm of life and the realm of death to permit a temporally broader discussion of structural violence. Although in death individuals are no longer experiencing bodies, they can maintain spiritual or symbolic power that shapes the behaviors of the living (Hallam et al. 1999; Nystrom 2017). Likewise, factors from biological, social, and cultural environments continue to impact these bodies even after death.

Embodiment refers to the body's incorporation of environmental influences, including biological, ecological, and social circumstances (Joyce 2000; Krieger 2005). In contemporary populations, embodiment has been used to discuss health in relation to sociopolitical processes and structural inequalities (e.g., Gravlee 2009; Krieger 2005). In bioarchaeology, the term "embodiment" is used to describe the biological (typically, skeletal) effects of social and cultural environments (e.g., Buikstra and Scott 2009; Joyce 2000). The contemporary use of the concept of embodiment with its strong emphasis on political-economic processes is compatible with studies that employ a biocultural framework to structure interpretations of past human health (Goodman 2013; Goodman and Leatherman 1998; Goodman et al. 1988), but it has also been employed to describe the physical body as a site of individual and community agency and sociopolitical display (Blom 2005; Duncan and Hofling 2011; Joyce 2000; Meskell 2000; Sofaer 2006). Notably, Nystrom (2011) used embodiment to describe the effects of political-economic processes that systematically denied resources to certain groups, and to structure interpretations about the sociopolitical role of the physical body in the context of postmortem examination (specifically, anatomical dissection).

Postmortem agency refers to the idea that dead bodies can engage and influence the behavior of the living (Crandall and Martin 2014). Theories of agency derive from several sources (e.g., Bourdieu 1977; Giddens 1984) and their use in archaeology remains contested (e.g., Dobres and Robb 2000; Dornan 2002). However, bioarchaeological studies of postmortem agency (Crandall and Harrod 2014; Tung 2014) have employed a more flexible definition, drawn from Gell (1998) and Robb (2004), who define agency on two axes. In the first type of agency, humans act on their own (primary or conscious agency); in the second, objects extend the agency of persons (secondary or effective agency). The latter resonates in studies of postmortem agency, which examine the roles of the dead in shaping the behaviors and relationships among living people (Betsinger and Scott 2014; Crandall and Harrod 2014; Sorensen 2009; Tung 2014).

The term "disembodiment" has been used to emphasize the process of dividing the body and reducing a person to a material object, and it is particularly applicable in the context of anatomical dissection and skeletal collections (de la Cova 2020; Nystrom 2014). Here, I focus on the personhood of an individual in their disembodied state, absent the (whole) body. A person can persist in constituent parts of a body, or even in symbolic representations of that individual (Geller 2004; Hallam et al. 1999; Lans 2018; Novak 2014). In the last few decades, many archaeologists have explored the role of divided bodies and body parts in societies and social relationships (e.g., Chapman 2000; Fowler 2001, 2002; Geller 2012; Rebay-Salisbury et al. 2010). The concept of disembodiment also reflects a particular anxiety that nineteenth-century individuals felt about the body and its meaning after death—namely, that something of the self persisted after death, and poor treatment of the body was a spiritual insult. For example, the same 1845 complaint referenced earlier also expressed concern over the fate of the deceased at the almshouse: "And may not this anxiety to have the remains cared for and protected after death be partly produced by the idea that the spirit may continue to be cognizant of what is done to the mortal part? *That death does not mean a total disconnection?*" (Lawrence 1905:161).

Excavation and Analysis of the Point San Jose Assemblage

In October of 2010, a skeletal assemblage was discovered in Fort Mason, formerly Point San Jose. The assemblage was described as a medical waste pit, because commingled human remains were deposited with mostly broken bottles and faunal remains (Gavette and Willey, this volume). Osteological analyses indicated these individuals were likely dissected, or at minimum used for medical training or display (Peters et al., this volume). Artifact analyses dated the pit between 1860 and 1900 (Locke Barton, this volume). Historical research indicated that physician Edwin Bentley, stationed at Point San Jose between 1869 and 1877, had a reputation as a collector of curiosities and was likely responsible for the dissection and disposal of the remains found at Point San Jose (Gavette, this volume). Bentley published case studies prolifically and was known for preserving human specimens for analysis; he also taught anatomy at the Medical College of the Pacific during this period (Gavette, this volume).

During the excavation, bones were removed and sorted by element, therefore any burial context specific to individual decedents was lost and the assemblage was almost entirely commingled. The Golden Gate National Recreational Area Conservancy contracted California State University (CSU), Chico, to complete the bioarchaeological analysis of the Point San Jose assemblage in 2016. All 4,450 bones and bone fragments were recorded, labeled using an adhesive, and photographed. As described in this volume, analyses of demography, paleopathology, and taphonomy were conducted and samples were removed for destructive isotopic analyses. At the time of this writing, the remains are sorted by element and stored in padded drawers inside storage cabinets at CSU, Chico.

Structural Violence at Point San Jose

The remains of the individuals interred at Point San Jose embodied the structural inequalities that those individuals experienced in life and that persisted after death. That is, not only is the structural violence they experienced during their lifetimes reflected in their biology, but the inequalities that shaped their death histories also had physical effects on their remains. Skeletal analyses indicated that these individuals were not members of the military but rather persons of diverse backgrounds, including males and females from various age groups, some of whom may have belonged to minoritized groups (Sgheiza and Willey, this volume). Evaluations of paleopathology indicated that some of these individuals experienced antemortem physical trauma, as well as degenerative changes and numerous nonspecific pathological responses (Milligan et al., this volume). Isotopic analyses indicated that six of the eight individuals evaluated for strontium and oxygen were likely not local to the Bay Area in the decade prior to their deaths (Bartelink and Hall, this volume). Evidence of postmortem examination (Peters et al., this volume) supports the inference that individuals at Point San Jose experienced structural inequality, because they received postmortem treatment reserved for the poorest and most marginalized groups. The presence of saw and incised cut marks suggests that the skeletons at Point San Jose were likely part of an anatomical and/or pathological collection (Dittmar and Mitchell 2015; Walker et al. 2014). Historical and osteological evidence indicates that members of lower socioeconomic classes and marginalized groups were disproportionately used as cadavers (Blakely and Harrington 1997; Davidson 2007; Harrington and Blakely 1995a, 1995b; Humphrey 1973; Nystrom 2017; Sappol 2002; Savitt 1982; Schultz 1992). Analyses from Point San Jose support not only the presence of partible postmortem practices but also fit the general demographic expectations for individuals subjected to dissection. Additionally, the individuals in the assemblage were not properly interred after dissection as mandated by law but rather discarded in a pit along with faunal remains and medical waste. It is, therefore, likely that individuals at Point San Jose were of low socioeconomic status and some may have been from minoritized groups.

The nature of this assemblage highlights the intentional fragmentation and curation of bodies and body parts in the nineteenth century by anatomists and surgeons in the US. This example of body partibility diverges somewhat from traditional partibility studies, which typically focus on partible practices as a form of ancestor veneration or in the context of intentional desecration or violation (e.g., Geller 2014; Mock 1998). The principal purpose of dissection during the nineteenth century was not desecration but rather medical training. Nonetheless, it had an explicit secondary purpose as punishment and even deterrence against indigence (Richardson 2000; Sappol 2002). In these cases, the body was objectified simultaneously as part collectors' item, part waste—parts of the body were considered valuable, while the rest was discarded.

Dissection as a process of objectification also introduces the concept of disembodiment at Point San Jose. Dissection divided whole bodies, and deposition commingled them and denied them the proper burials they were legally due. In many ways, these partible practices symbolically divorced the dead from their living social identities through objectification (Sappol 2002). In fact, dissected bodies were so thoroughly objectified that they were not even considered property and therefore could not technically be stolen or even owned, legally speaking (Richardson 2000). And yet, this same division simultaneously reinforced those social identities and the identities of others who shared them in a kind of anatomical synecdoche: dissected parts of (dead) bodies signaled that the whole (living) bodies deserved dissection.

Reflection on the death histories of the individuals in this assemblage provides a strong case for the presence of postmortem agency and provokes reflection on the role of bioarchaeological practice in the cycle of structural violence. The disembodied individuals represented at Point San Jose not only shaped the behaviors of the individuals involved in their postmortem examination and disposal but continued to shape bioarchaeologists' behaviors today. Precisely because these individuals were commingled and showed evidence of dissection, they were of interest to bioarchaeologists (Willey et al. 2016, 2018). Consequently, these remains were excavated (and in the process, further commingled); they have been subjected to photography, microscopy, radiography, and x-ray fluorescence spectrometry; and they have been destructively sampled for isotopic analysis. Although bioarchaeologists have curated these remains in an attempt to preserve them, these individuals remain parted and have yet to experience the burial they were guaranteed by law.

Although all of these analyses were conducted to learn more about this marginalized group, it is vitally important that bioarchaeologists consider our role as potential perpetrators of structural violence against marginalized individuals (Watkins 2018). Bioarchaeologists do not discover history but actively construct, shape, and re-shape narratives and meanings of individuals and social relationships in the past (Geller and Suri 2014). Bioarchaeologists have the opportunity to learn about systems of power from the remains of marginalized bodies, information that is seldom recorded from people who were systematically silenced. Zuckerman et al. (2014) explicitly addressed this, arguing that the bioarchaeological analysis of marginalized groups is ethically justifiable only under the condition that researchers study that marginalization. That is, if skeletal data from marginalized groups are to be collected, they must be sufficiently contextualized to allow for interpretations of the effects of power and oppression and to illuminate the narratives that are too often absent from historical records (Zuckerman et al. 2014). Embracing embodiment/disembodiment and theories of postmortem agency implies a recognition that dead bodies are not entirely dead. As Arnold (2014:523) put it, they are only mostly dead, and "mostly dead is slightly alive." The assemblage at Point San Jose represents a unique window into the lives and deaths of marginalized persons in nineteenth-century San Francisco, but this knowledge came at great cost to the individuals in the assemblage. It is therefore imperative that we bioarchaeologists explicitly recognize our active role in making meaning from the dead and our obligations to those we study.

Acknowledgments

Many thanks to P. Willey, Eric Bartelink, Colleen Milligan, and Peter Gavette for inviting me to contribute to this volume. I am also very grateful to Kelly Knudson, Aliya Hoff, Brooke Hundtoft, Valerie Leah, Jessica Rothwell, and Jaclyn Thomas for thought-provoking discussions and valuable feedback on early drafts of this chapter.

References Cited

Arnold, Bettina
2014 Life after Life: Bioarchaeology and Post-Mortem Agency. *Cambridge Archaeological Journal* 24(3):523–529.

Berry, Daina Ramey
2017 *The Price for Their Pound of Flesh: The Value of the Enslaved, from Womb to Grave, in the Building of a Nation.* Beacon Press, Boston.

Betsinger, Tracy K., and Amy B. Scott
2014 Governing from the Grave: Vampire Burials and Social Order in Post-Medieval Poland. *Cambridge Archaeological Journal* 24(3):467–476.

Blakely, Robert L., and Judith M. Harrington (editors)
1997 *Bones in the Basement: Postmortem Racism in Nineteenth-Century Medical Training.* Smithsonian Institution Press, Washington, DC.

Blakey, Michael L.
1998 Beyond European Enlightenment: Toward a Critical and Humanistic Human Biology. In *Building a New Biocultural Synthesis: Political-Economic Perspectives on Human Biology,* edited by A. H. Goodman and T. Leatherman, pp. 379–406. University of Michigan Press, Ann Arbor.

Blom, Deborah E.
2005 Embodying Borders: Human Body Modification and Diversity in Tiwanaku Society. *Journal of Anthropological Archaeology* 24(1):1–24.

Bourdieu, Pierre
1977 *Outline of a Theory of Practice.* Cambridge University Press, Cambridge.

Bourgois, Phillipe
2009 Recognizing Invisible Violence: A Thirty-Year Ethnographic Retrospective. In *Global Health in Times of Violence,* edited by B. Rylko-Bauer, L. Whiteford, and P. Farmer, pp. 17–40. School for Advanced Research Press, Santa Fe, New Mexico.

Bright, Lisa N.
2020 Structural Violence: Epistemological Considerations for Bioarchaeology. In *Theoretical Approaches in Bioarchaeology,* edited by C. M. Cheverko, J. Prince-Buitenhuys, and M. Hubbe, pp. 131–149. Taylor and Francis Group.

Buikstra, Jane E., and Rachel E. Scott
2009 Key Concepts in Identity Studies. In *Bioarchaeology and Identity in the Ameri-*

cas, edited by K. J. Knudson and C. M. Stojanowski, pp. 24–55. University Press of Florida, Gainesville.

Chapman, John C.

2000 *Fragmentation in Archaeology: People, Places and Broken Objects in the Prehistory of South-Eastern Europe.* Routledge, London.

Crandall, John J., and Ryan P. Harrod

2014 Ghostly Gunslingers: The Postmortem Lives of the Kiel Brothers, Nevada's First Frontiersmen. *Cambridge Archaeological Journal* 24(3):487–497.

Crandall, John J., and Debra L. Martin

2014 Special Section: The Bioarchaeology of Postmortem Agency (Guest Editors: John J. Crandall and Debra L. Martin) The Bioarchaeology of Postmortem Agency: Integrating Archaeological Theory with Human Skeletal Remains. *Cambridge Archaeological Journal* 24(3):429–435.

Crossland, Zoe

2009 Acts of Estrangement: The Post-Mortem Making of Self and Other. *Archaeological Dialogues* 16(1):102–125.

Davidson, James M.

2007 "Resurrection Men" in Dallas: The Illegal Use of Black Bodies as Medical Cadavers (1900–1907). *International Journal of Historical Archaeology* 11:193–220.

de la Cova, Carlina

2019 Marginalized Bodies and the Construction of the Robert J. Terry Anatomical Skeletal Collection. In *Bioarchaeology of Marginalized People,* edited by M. Mant and A. Holland, pp. 133–155. Elsevier, Inc.

2020 Making Silenced Voices Speak: Restoring Neglected and Ignored Identities in Anatomical Collections. In *Theoretical Approaches in Bioarchaeology,* edited by C. M. Cheverko, J. Prince-Buitenhuys, and M. Hubbe, pp. 150–169. Taylor and Francis Group.

Dittmar, Jenna M., and Piers D. Mitchell

2015 A New Method for Identifying and Differentiating Human Dissection and Autopsy in Archaeological Human Skeletal Remains. *Journal of Archaeological Science: Reports* 3:73–79.

Dobres, Marcia-Anne, and John E. Robb

2000 Agency in Archaeology: Paradigm or Platitude? In *Agency in Archaeology,* edited by M. Dobres and J. E. Robb, pp. 3–17. Routledge, New York.

Dornan, Jennifer L.

2002 Agency and Archaeology: Past, Present, and Future Directions. *Archaeological Method and Theory* 9(4):303–329.

Duncan, William N., and Charles Andrew Hofling

2011 Why the Head? Cranial Modification as Protection and Ensoulment among the Maya. *Ancient Mesoamerica* 22(1):199–210.

Farmer, Paul

1996 On Suffering and Structural Violence: A View from Below. *Daedalus* 125(1):251–283.

2003 *Pathologies of Power: Health, Human Rights, and the New War on the Poor.* University of California Press, Berkeley.

2004 An Anthropology of Structural Violence. *Current Anthropology* 45:305–325.

Fowler, Christopher

2001 Personhood and Social Relations in the British Neolithic with a Case Study from the Isle of Man. *Journal of Material Culture* 6:137–163.

2002 Body Parts: Personhood and Materiality in the Manx Neolithic. In *Thinking through the Body: Archaeologies of Corporeality,* edited by Y. Hamilakis, M. Pluciennik, and S. Tarlow, pp.47–69. Kluwer Academic/Plenum Publishers, London.

Galtung, Johan

1969 Violence, Peace, and Peace Research. *Journal of Peace Research* 6(3):167–191.

1975 Is Peaceful Research Possible: On the Methodology of Peace Research. In *Peace, Research, Education, Action: Essays in Peace Research,* edited by J. Galtung, Vol. 1, pp. 263–79. Christian Ejlers, Copenhagen.

1990 Cultural Violence. *Journal of Peace Research* 27(3):291–305.

Garment, Ann, Susan Lederer, Naomi Rogers, and Lisa Boult

2007 Let the Dead Teach the Living : The Rise of Body Bequeathal in 20th-Century America. *Academic Medicine* 82(10):1000–1005.

Gell, Alfred

1998 *Art and Agency: An Anthropological Theory.* Clarendon Press, Oxford.

Geller, Pamela L.

2004 *Transforming Bodies, Transforming Identities: A Consideration of Pre-Columbian Maya Corporeal Beliefs and Practices.* University of Pennsylvania Press, Philadelphia.

2012 Parting (with) the Dead: Body Partibility as Evidence of Commoner Ancestor Veneration. *Ancient Mesoamerica* 23(1):115–130.

2014 Sedimenting Social Identity: The Practice of Pre-Columbian Maya Body Partibility. In *The Bioarchaeology of Space and Place,* edited by G. Wrobel, pp. 15–38. Springer, New York.

Geller, Pamela L., and Miranda Stockett Suri

2014 Relationality, Corporeality and Bioarchaeology: Bodies qua Bodies, Bodies in Context. *Cambridge Archaeological Journal* 24(3):499–512.

Giddens, Anthony

1984 *The Constitution of Society: Outline of a Theory of Structuration.* University of California Press, Berkeley.

Goodman, Alan H.

2013 Bringing Culture into Human Biology and Biology Back into Anthropology. *American Anthropologist* 115(3):359–373.

Goodman Alan H., and Thomas L. Leatherman

1998 *Building a New Biocultural Synthesis: Political-Economic Perspectives on Human Biology.* University of Michigan Press, Ann Arbor.

Goodman, Alan H., R. Brooke Thomas, Alan C. Swedlund, and George J. Armelagos

1988 Biocultural Perspectives on Stress in Prehistoric, Historical, and Contemporary Population Research. *American Journal of Physical Anthropology* 31(S9):169–202.

Gravlee, Clarence C.

2009 How Race Becomes Biology : Embodiment of Social Inequality. *American Journal of Physical Anthropology.* 57:47–57.

Hallam, Elizabeth, Jenny Hockey, and Glennys Howarth

1999 *Beyond the Body: Death and Social Identity.* Routledge, London.

Harrington, Judith M., and Robert L. Blakely

1995a Bones in the Basement: Bioarchaeology of Historic Remains in Nonmortuary Contexts. In *Bodies of Evidence: Reconstructing History through Skeletal Analysis,* edited by A. L. Grauer, pp. 105–120. Wiley-Liss, New York.

1995b Rich Man, Poor Man, Beggar Man, Thief: The Selectivity Exercised by Graverobbers at the Medical College of Georgia, 1837–1887. In *Grave Reflections: Portraying the Past through Cemetery Studies,* edited by S. R. Saunders and A. Herring, pp. 153–178. Canadian Scholars, Toronto.

Harrod, Ryan P., Jennifer L. Thompson, and Debra L. Martin

2012 Hard Labor and Hostile Encounters: What Human Remains Reveal about Institutional Violence and Chinese Immigrants Living in Carlin, Nevada (1885–1923). *Historical Archaeology* 46(4):85–111.

Humphrey, David C.

1973 Dissection and Discrimination: The Social Origins of Cadavers in America, 1760–1915. *Bulletin of the New York Academy of Medicine* 49:819–825.

Joyce, Rosemary A.

2000 Archaeology of the Body. *Annual Review of Anthropology* 34(1):139–158.

Klaus, Haagen D.

2012 The Bioarchaeology of Structural Violence: A Theoretical Model and a Case Study. In *The Bioarchaeology of Violence,* edited by D. Martin, R. Harrod, and V. Pérez, University of Florida Press.

Köhler, Gernot, and Norman Alcock

1976 An Empirical Table of Structural Violence. *Journal of Peace Research* 13:343–356.

Krieger, Nancy

2005 Embodiment: A Conceptual Glossary for Epidemiology. *Journal of Epidemiology and Community Health* 59(5):350–355.

Lans, Aja

2018 "Whatever Was Once Associated with Him, Continues to Bear His Stamp": Articulating and Dissecting George S. Huntington and His Anatomical Collection. In *Bioarchaeological Analyses and Bodies: New Ways of Knowing Anatomical and Archaeological Skeletal Collections,* edited by P. K. Stone, pp. 11–26. Springer International, Cham, Switzerland.

Lawrence, Charles

1905 *History of the Philadelphia Almshouses and Hospitals.* Princeton University, Princeton, New Jersey.

Marmot, Michael, Sharon Friel, Ruth Bell, Tanja AJ Houweling, and Sebastian Taylor

2008 Closing the Gap in a Generation: Health Equity through Action on the Social Determinants of Health. *The Lancet* 372(9650):1661–1669.

Martin, Debra L., Nancy J. Akins, Bradley J. Crenshaw, and Pamela K. Stone
2008 Inscribed in the Body, Written on the Bones: The Consequences of Social Violence at La Plata. In *Social Violence in the Prehispanic American Southwest,* edited by D. L. Nichols and P. L. Crown, pp. 98–122. University of Arizona Press, Tucson.
Meskell, Lynn
2000 Cycles of Life and Death: Narrative Homology and Archaeological Realities. *World Archaeology* 31(3):423–441.
Mock, Shirley B.
1998 The Defaced and the Forgotten: Decapitation and the Flaying/Mutilation as a Termination Event at Colha, Belize. In *The Sowing and the Dawning: Termination, Dedication, and Transformation in the Archaeological and Ethnographic Record of Mesoamerica,* edited by S. B. Mock, pp. 113–124. University of New Mexico Press, Albuquerque.
Nelkin, Dorothy, and Lori Andrews
1998 Do the Dead Have Interests? Policy Issues for Research after Life. *American Journal of Law and Medicine.* 24(2–3):261–291.
Novak, Shannon A.
2014 Leave Taking: Materialities of Moving over Land. *Cambridge Archaeological Journal* 24(3):477–485.
Nystrom, Kenneth C.
2011 Postmortem Examinations and the Embodiment of Inequality in 19th Century United States. *International Journal of Paleopathology* 1:164–172.
2014 The Bioarchaeology of Structural Violence and Dissection in the 19th-Century United States. *American Anthropologist* 116(4):765–779.
2017 *The Bioarchaeology of Dissection and Autopsy in the United States.* Springer International Publishing, Cham, Switzerland.
Quesada, James, Laurie Kain Hart, and Phillippe Bourgois
2011 Structural Vulnerability and Health: Latino Migrant Laborers in the United States. *Medical Anthropology* 30(4):339–362.
Rebay-Salisbury, Katharina, Marie Louise S. Sørensen, and Jessica Hughes
2010 *Body Parts and Bodies Whole: Changing Relations and Meanings.* Oxbow Books, Oxford, United Kingdom.
Richardson, Ruth
2000 *Death, Dissection, and the Destitute.* University of Chicago Press, Chicago.
Richardson, Ruth, and Brian Hurwitz
1987 Medical History Jeremy Bentham's Self Image: An Exemplary Bequest for Dissection. *British Medical Journal* 295:195–198.
Robb, John E.
2004 The Extended Artefact and the Monumental Economy: A Methodology for Material Agency. In *Rethinking Materiality: The Engagement of Mind with the Material World,* edited by E. DeMarrais, C. Gosden, and C. Renfew, pp 131–140. McDonald Institute for Archaeological Research, Cambridge.
Rylko-Bauer, Barbara, and Paul Farmer
2016 Structural Violence, Poverty, and Social Suffering. In *The Oxford Handbook of*

the Social Science of Poverty, edited by D. Brady and L. M. Burton, pp. 47–81. Oxford University Press, Oxford, United Kingdom.

Sappol, Michael

2002 *A Traffic of Dead Bodies: Anatomy and Embodied Social Identity in Nineteenth-Century America.* Princeton University Press, Princeton, New Jersey.

2017 Foreword II: The Afterlife of Death. In *The Bioarchaeology of Dissection and Autopsy in the United States,* edited by K. C. Nystrom, pp. vi-xi. Springer International, Cham, Switzerland.

Savitt, Todd L.

1982 The Use of Blacks for Medical Experimentation and Demonstration in the Old South. *Journal of Southern History* 48:331–348.

Scheper-Hughes, Nancy

1992 *Death without Weeping: The Violence of Everyday Life in Brazil.* University of California Press, Berkeley.

Scheper-Hughes, Nancy, and P. Bourgois

2004 *Violence in War and Peace: An Anthology.* Blackwell, Malden, Massachusetts.

Schultz, Suzanne M.

1992 *Body Snatching: The Robbing of Graves for the Education of Physicians in Early Nineteenth Century America.* McFarland and Company, Jefferson, North Carolina.

Singer, Merrill

2009 *Introduction to Syndemics: A Critical Systems Approach to Public and Community Health.* Jossey-Bass, San Francisco.

Singer, Merrill, Nicola Bulled, Bayla Ostrach, and Emily Mendenhall

2017 Syndemics and the Biosocial Conception of Health. *The Lancet* 389(10072): 941–950.

Singer, Merrill, and Scott Clair

2003 Syndemics and Public Health: Reconceptualizing Disease in Bio-Social Context. *Medical Anthropology Quarterly* 17(4):423–441.

Sofaer, Joan

2006 *The Body as Material Culture.* Cambridge University Press, Cambridge, United Kingdom.

Sorensen, Tim Flohr

2009 The Presence of the Dead: Cemeteries, Cremation, and the Staging of Non-Place. *Journal of Social Archaeology* 9(1):110–135.

Southwood Smith, Thomas

1824 Use of the Dead to the Living. *Westminster Review* 2:59–97.

Tremblay, Lori A., and Sara Reedy

2020 *The Bioarchaeology of Structural Violence: A Theoretical Framework for Industrial Era Inequality.* Springer.

Tung, Tiffiny A.

2014 Agency, 'Til Death Do Us Part? Inquiring about the Agency of Dead Bodies from the Ancient Andes. *Cambridge Archaeological Journal* 24(3):437–452.

Walker, Don, Natasha Powers, and Louise Fowler
2014 Resurrection: Who Is It Good For? The Price of Achievement at the London Hospital. *Post-Medieval Archaeology* 48(2):388–397.

Watkins, Rachel J.
2018 The Fate of Anatomical Collections in the US: Bioanthropological Investigations of Structural Violence. In *Identified Skeletal Collections: The Testing Ground of Anthropology?* edited by C. Y. Henderson and F. Alves Cardoso, pp. 169–186. Archaeopress, London.

Willey, P., Colleen Milligan, Eric Bartelink, Maria Cox, Kristin Broehl, Jessica, Curry, and Valerie Sgheiza
2016 *Descriptive Analysis of Human Skeletal Material from Point San Jose (Fort Mason), Golden Gate National Recreation Area.* Submitted to Golden Gate National Recreation Area Conservancy. Proposal 16–13[1]. Copies available from Golden Gate National Recreation Area.

Willey, P., Eric Bartelink, Colleen Milligan, Maria Cox, Mallory Peters, Kasey Cole, Jessica Curry, Sarah Hall, Kelsie Hart, Sam Mijal, Julia Prince-Buitenhuys, and Valerie Sgheiza
2018 *Osteological Analyses of a Medical Waste Pit from Point San Jose (Fort Mason), Golden Gate National Recreation Area.* Submitted to Golden Gate National Recreation Area Conservancy. Proposal 16–13[2]. Copies available from Golden Gate National Recreation Area.

Zuckerman, Molly K., Kelly R. Kamnikar, and Sarah A. Mathena
2014 Recovering the "Body Politic": A Relational Ethics of Meaning for Bioarchaeology. *Cambridge Archaeological Journal* 24(3):513–522.

16

Conclusions and Future Research

ERIC J. BARTELINK, PETER GAVETTE, P. WILLEY, AND COLLEEN F. MILLIGAN

The medical waste deposit discovered behind the former hospital at Point San Jose provides a unique window into late nineteenth-century anatomical dissection practices in the San Francisco Bay Area as well as the social context in which bodies were acquired for postmortem study. Although there have been other discoveries of nineteenth-century deposits containing anatomized body parts reported in the literature (Introduction, this volume), only one other is from the Pacific Coast and no others are associated with a military hospital. The unexpected discovery of the deposit raised many questions, some of which are answered through this study.

The chapters in this volume contribute important information on the discovery and archaeological context of the medical waste deposit. In addition, these contributions provide the historical foundations of the military presence at Point San Jose and the Bay Area, as well as medical education and anatomical practice during the late nineteenth century. The osteological and isotopic studies establish the number of individuals and demographic characteristics likely represented by 4,550 skeletal elements, the prevalence of stress and disease indicators, and the reconstruction of diet and origin of individuals interred in the deposit. Other contributions provide historical context on how physicians obtained medical specimens for dissection, the most likely surgeon involved in the acquisition and disposal of the remains, and the history of cadaver acquisition for medical training. The final paper provides a social bioarchaeological perspective that views anatomical dissection through the lens of structural violence. These chapters offer descriptive, historical, and social theoretical perspectives on the lives and deaths of the individuals interred in the pit, as well as the archaeological context of the deposit. As other similar archaeological discoveries, the Point San Jose medical waste deposit reflects human behavior—in this case, the transformation of corpses into body parts (i.e., partible persons) by the physicians and their students.

In the late nineteenth century, the use of unclaimed bodies for anatomical dissection, surgical training, and teaching reflected the growing need for cadavers to train the next generation of physicians (Sappol 2002). Given the liberal laws in California regarding the use of unclaimed bodies for dissection at that time, physicians took advantage of the ready supply of cadavers from local almshouses, hospitals, poorhouses, county jails, and state prisons, minimizing the need for grave robbing. Further, the use of unclaimed bodies for dissection, often those of paupers, was justified on the basis that these individuals owed a debt to society due the social services provided to them in life (Sappol 2002). These anatomized Point San Jose individuals likely reflected this practice. However, why these particular remains were buried behind the post hospital at Point San Jose instead of being interred in a cemetery following their postmortem study is less clear. Therefore, at the outset of this project, we sought to identify the purpose of the medical waste deposit found behind the hospital.

Here we reexamine the six broad questions posed in the introductory chapter.

1) What was the purpose of the anatomical and medical waste deposit at Point San Jose? Did the remains represent a pathological collection, cadavers, autopsied corpses, or members of the military?

The remains interred in the pit show evidence of postmortem study most consistent with nineteenth-century anatomical dissection practices. Although some skeletal modifications are also consistent with autopsy (e.g., circumferential craniotomies, saw cuts through the sterna and clavicles), numerous incised cut marks throughout the body indicate dissection practices aimed at reflecting soft tissues and exposing deeper structures of the body (Peters et al., this volume). In addition, several skeletal elements show evidence of saw cuts, consistent with prosection (e.g., hemi-mandibles) or with surgical training, possibly relating to amputation (e.g., bisected long bones). None of the sharp-force defects are associated with surgical intervention, and therefore suggest postmortem examination. As a supportive observation, some evidence for animal dissection may have been present (Cole et al., this volume). Although some human skeletal elements and dental remains exhibit evidence of pathological conditions (e.g., degenerative changes, fractures, periosteal reaction, dental caries, enamel hypoplasia), the overall prevalence was lower than would be expected if the remains represented the contents of a pathological curio collection (Milligan et al., this volume). Considering some of the body parts buried in the deposit showed articulated segments (e.g., thoraxes, vertebral columns, and pelvic girdles), some of the skeletal elements must have been interred with soft tissue still attached (Gavette and Willey, this volume). Further, the biological profile information is inconsistent with a solely military sample based on the short estimated statures (most below military enlistment standards; Willey, this volume), the presence of females, subadults and some older adults (Sgheiza and Willey, this volume), and the unusual disposal context (Gavette and Willey, this volume).

2) When were the bones and artifacts buried in the pit at Point San Jose and what did their disposal represent?

The dating of the Point San Jose assemblage is based primarily on historical artifacts interred in the pit. Because the pit appears to represent a single deposition instead of an accumulation of materials over time, the artifacts interred with the remains provide the strongest basis for dating the deposit. The date of the artifact assemblage, which was dominated by medical bottles and food and beverage containers, suggests a *terminus post quem* after the Civil War, most likely during the 1870s (Locke Barton, this volume).

Discarding the bones and artifacts represented a purposeful and intentional act, not an incidental or haphazard one. A steep-sided pit was dug, and human bones and body segments were placed in it first. The event emphasized the human elements with the artifacts and faunal elements being of secondary importance (Gavette and Willey, this volume). The later part of the nineteenth century saw medical education and medical practice becoming more aligned with the scientific study of the mechanisms of the human body that are common to all individuals. It is possible that the discarded human remains behind the Point San Jose hospital during the post–Civil War period are evidence of a change in medical understanding that emphasized a more universalistic and empirical approach (Collison 2015). More likely, however, the deposit probably occurred as a matter of convenience, decorum, and hygiene following conclusion of their anatomical study and dissection.

3) Did the human remains from Point San Jose include all elements of all skeletons? Or did they represent selected skeletal elements and body segments?

The remains interred in the deposit did not consist of intact, whole individuals. The age-based minimum number of individuals indicated at least 25 individuals are represented, although the actual number could be as high as 40 individuals (Cox et al., this volume). Overall recovery rates of skeletal elements were low, especially for skulls and femora. Distal elements of limbs were underrepresented while proximal portions were overrepresented. Thus, some elements and body portions were underrepresented in the sample most likely from selective retention and curation or disposal in another manner. Attempts to rearticulate disassociated elements proved ineffective. Few disarticulated skeletal elements could accurately be matched by elemental analyses, joint articulation, or pair-matching.

4) What is the demographic composition of the human remains? What sex, age, and ancestry groups do the Point San Jose skeletons represent? Were these individuals local or nonlocal to the San Francisco Bay Area?

The demographic composition of the Point San Jose series includes males and females of various ages and ancestral groups (Sgheiza and Willey, this volume). Interestingly, while most remains are middle-aged adults, fetal remains are also present. The Point San Jose age distribution is inconsistent with either a military sample or the 1870 California census and mortality data. Both males and females are represented, with no major bias for one sex. Although ancestry estimations are limited to few individuals and thwart meaningful comparisons with census data, some of the Point San Jose individuals show evidence of representing disadvantaged groups. These data suggest the possible bias toward minority groups, likely representing individuals of low socioeconomic status (SES). However, unlike many other similar skeletal series, few if any individuals of African ancestry were recognized in the Point San Jose assemblage. These assessments suggest that ancestry or ethnicity may not have been the key factor in cadaver selection, with regionally available lower SES possibly playing a larger role in who was dissected.

Stable carbon and nitrogen isotope data of individuals from Point San Jose suggest that regional patterns dominated foodways (Hall et al., this volume). Overall, dietary variation was minimal and similar to individuals from a partially contemporaneous pauper cemetery located in Santa Clara County (Valley Medical Center). Low carbon isotope values coupled with moderately high nitrogen values suggest a dietary focus on terrestrial plants and animals from C_3 ecosystems as well as freshwater fish. Bay resources, such as marine and estuarine fish, appear to have been of little importance in their diet. Comparisons between California and other regions of the US show differences and reflect regional patterning in foodways independent of social status.

Attempting to determine childhood and adulthood residence patterns proved difficult. Oxygen and strontium isotope data from eight Point San Jose individuals suggest that the anatomized individuals likely lived outside of the Bay Area during their childhood and adult years (Bartelink and Hall, this volume). Individuals who died soon after arriving in San Francisco may have been among those claimed for dissection if they were of low SES (e.g., paupers) and were unclaimed 24 hours after death. However, we cannot exclude other possibilities, such as cadavers being transported to San Francisco from neighboring counties.

5) What do the human remains indicate about health and stress during the people's lives?

We expected that the remains from the medical waste deposit at Point San Jose would represent individuals of low SES and that they would demonstrate indications of stress associated with such status. However, the overall prevalence of paleopathological conditions is low, with only 20 percent of observed skeletal material showing evidence of pathological conditions (Milligan et al., this volume). New bone formation is the most common condition observed, most often consisting of degenerative disease of the appendicular joints and spine. Dental caries rates are comparable to or lower than other nineteenth-century skeletal series, although linear enamel hypoplasia defects were especially common, affecting 19.4 percent of all teeth examined. The stature estimation mean stood about midway when compared with worldwide data, and when sexes were estimated, the Point San Jose females averaged the same as some other nineteenth-century women. These data suggest a moderate level of stress among individuals in our sample, but the evidence is limited by the commingled nature of the assemblage.

6) On a broader level, what role did the Point San Jose deposit represent in post–Civil War medicine and military medicine? And more specifically, who commanded gathering, use, and disposal of the human remains?

The dating of the medical waste deposit based on artifacts coincides with Dr. Edwin Bentley's posting at the hospital from 1869–1875. Bentley had prominent roles as a professor of Descriptive and Microscopic Anatomy and Pathology for the Medical College of the Pacific, as a surgeon for the hospital at Point San Jose, the Presidio, and as official pathologist for the City and County Hospital of San Francisco (Gavette, this volume). He published numerous papers on autopsies, pathological findings, and submitted both soft tissue and skeletal specimens to museums for curation (Spatola and Pearlstein, this volume). Bentley's ongoing involvement with anatomical teaching, pathology, research, and collection of specimens for museums makes him the most likely person responsible for the acquisition of the remains. The year before his post began at Point San Jose in 1870, he was accused of retaining body parts from amputations at his former home in Norwich, Connecticut. In 1878, three years after leaving his post at Point San Jose while at the Army's Little Rock Barracks, he was accused of dissecting the remains of bodies obtained from the local city hospital and burying the body parts near the dead house. Considering this evidence, it is likely that Bentley was directly or indirectly responsible for the anatomized remains that were interred in the pit behind the hospital at Point San Jose (Gavette, this volume). Although we were unable to find a historical record of a Point San Jose dead house, it may have been associated with the hospital, providing a place where bodies could be dissected out of the public eye. California anatomy laws at that time indicated that cadavers used in postmortem studies were to be interred in a public cemetery after their use, which clearly did not happen in this case. However, as noted by Gavette and Willey (this volume), the remains were interred in a systematic fashion that may have reflected some level of respect.

The medical waste deposit at Point San Jose represents an era when the need for bodies in medical school dissection courses was growing. The liberal anatomy laws in California provided an ample supply of bodies for San Francisco physicians to train the next generation, but this resource almost exclusively relied on the unclaimed bodies of the unfortunates—paupers, indigents, and the people forgotten to history. In many ways, the analyses presented in this volume provide a voice to these individuals who were so unceremoniously disposed of at Point San Jose. As bioarchaeologists and archaeologists, we must be cognizant that our findings highlight systemic inequality of marginalized groups in the past (Hall, this volume).

Future Research

As most research, this work has raised as many questions as it answered. Answers to those recently emerging questions lie in further examining the remains and artifacts with new approaches. We consider additional topics that could be pursued, presented here in more or less the same sequence that they occur in the body of this work.

Archaeologically based topics come first. Cultural artifacts hold additional, unlocked information. As noted earlier (Locke Barton, this volume), artifacts from the Point San Jose pit represent a select and unusual assemblage. In particular, the large frequency of pharmaceutical bottles and few alcohol bottles proved extraordinary when compared with other historical hospital assemblages. Analyzing artifacts that have been recovered in other locations at Point San Jose other than the pit might show differences between normal trash that accumulates over time and the one-time event that the Point San Jose pit represents. The non-pit artifacts would enhance an understanding of the hospital's purposes and functions.

The archaeological context should be expanded and elaborated, particularly emphasizing specific elements. Special elements warranting further spatial investigation include human sawed limb bones and bones with incisions. Understanding where and how those most thoroughly anatomized remains were deposited in the pit would aid their interpretation. Those elements that merited extensive anatomical attention may have received special attention during their disposal. For similar reasons, placement of skulls in the pit should be pursued. Their placement might indicate if different ancestry groups and sexes were discarded in different manners. Finally, the context of some faunal elements requires further assessment. In particular, the cut dog remains deserve attention to see if they were viewed as mere trash, or on the other hand, may have found their place alongside the other, although human, dissected anatomical specimens. Locations in the Point San Jose pit may aid with interpreting the anatomized human and canine elements.

Similar to artifacts, faunal analyses should be expanded beyond those elements recovered in the Point San Jose deposit to include entire assemblages outside the pit. As the bottles, the pit fauna may represent selected species and elements when contrasted with those recovered outside the pit. In addition, the Point San Jose faunal assemblage may provide insights into ethnicity and cuisine. Based on preferences in animal taxa, meat cuts, butchery tools and preparation techniques, other historical faunal studies have inferred ethnicity and ethnic differences. Similar influences may be possible with the Point San Jose animal bones, although the diet results (Hall et al., this volume) failed to identify associations with other geographic regions within the US. In addition to ethnic diets, patients in the post hospital may have been fed different and specially prepared foods, possibly those considered a more nourishing and restorative diet, rather than standard rations. In addition, military ranks ate different foods, and rank differences in meats and cuts might be indicated by further faunal assessment.

In future research, we hope to narrow the region of origin for Point San Jose individuals by using additional isotopes. Isoscape prediction maps should also be expanded to include regions beyond the continental US, including Asia and Latin America. Used in conjunction with morphological ancestry assessments, these new isotope results could provide more reliable identifications of the heritages that these remains represent. Refined origin and ancestry estimates would also offer a better understanding of the selection of anatomical specimens for dissection in the nineteenth-century San Francisco Bay Area.

Future research should include a molecular approach. With substantial funding and effort, DNA testing could refine ancestral backgrounds and locations of origins as well as identify a more exact number of individuals present.

The current dietary data do not permit evaluations of the effects of local socioeconomic and sociopolitical impacts. With a larger nineteenth-century sample from the Bay Area that includes immigrants, locals, and all social classes, the impacts of SES on foodways may be better understood.

Two disease manifestations displayed high frequencies in the Point San Jose human materials: linear enamel hypoplasias and joint disease. Their study should be expanded to encompass demographic data, including age, sex, and ancestry if possible. In addition, more extensive microscopic studies might identify the type and extent of pathological defects observed.

Dental pathologies deserve further assessment. The present work shows that the Point San Jose dentitions have low caries rates contrasted with comparable nineteenth-century samples—a surprising finding. Does this low caries rate indicate better oral health or dentitions with greater antemortem tooth loss caused by previous caries? The Point San Jose dentitions should be examined for antemortem tooth loss and other dental conditions for comparisons with other samples.

A more thorough understanding of nineteenth-century medical education and anatomical practice is in order. This historical background to anatomical training could establish a framework for interpreting the distribution and direction of cut marks related to autopsy and dissection procedures. In addition, the nineteenth-century tools used in those processes could be identified. Establishing the historical background, dissection procedures, and the equipment used in dissection will prove valuable for more securely identifying the purpose of the Point San Jose assemblage (e.g., its potential uses for dissection, surgery practice, autopsy, teaching, etc.). In addition, more sophisticated photographic techniques, such as reflectance transformation imaging, might refine interpretations of cutting techniques and tools employed.

Similar historical research might identify the individuals whose cadavers were selected. Although the 1906 earthquake destroyed many of the original documents in San Francisco (see Gavette, this volume), some records escaped, and pre-earthquake published works do exist. The possibility of matching mortuary records (now in the San Francisco Department of Public Health) and the coroner records in the annual municipal reports and possibly other sources would prove a useful first step.

Ethical considerations for studying and curating or reinterring the Point San Jose collection should be explored. As part of this inquiry, recommendations for considerate handling and memorialization should be addressed.

Other bioarchaeology studies dealing with similar anatomical and autopsy assemblages have argued that disadvantaged groups were disproportionately selected for dissection (e.g., Nystrom 2011, 2017). Those instances of structural violence were based largely on social race and archaeological context. Eastern and southern Blacks were the frequent targets for dissection. If structural violence is used to interpret corpse selection for anatomical dissection, then with the Point San Jose results that explanation must be expanded to include additional disadvantaged groups.

Lower SES is the most likely alternate explanation. Additional efforts must be made to assess the SES based on skeletal materials at Point San Jose. Several related parameters have already been examined (diet, stature, linear enamel hypoplasia, dental caries) with mixed biological indications of lower SES. The pursuit should be expanded to include additional dental health issues and radiopaque transverse (Harris) lines. Further research on more affluent social groups from the Bay Area could help bolster the argument that the Point San Jose skeletal remains reflect the preferential selection of low SES individuals for dissection.

References

Collison, Patrick

2015 Overview of Nineteenth Century Medicine. In *Health of the Seventh Cavalry*, edited by P. Willey and Douglas D. Scott, pp. 43–62. University of Oklahoma Press, Norman.

Nystrom, Kenneth C.

2011 Postmortem Examinations and the Embodiment of Inequality in 19th Century United States. *International Journal of Paleopathology* 1(3-4):164–172.

Nystrom, Kenneth C., editor.

2017 *The Bioarchaeology of Dissection and Autopsy in the United States*. Springer International Publishing, Cham, Switzerland.

Sappol, Michael

2002 *A Traffic of Dead Bodies: Anatomy and Embodied Social Identity in Nineteenth-Century America*. Princeton University Press, Princeton, New Jersey.

Contributors

Eric J. Bartelink is professor of anthropology and codirector of the Human Identification Laboratory at California State University, Chico. He conducts bioarchaeological research on prehistoric and historic remains in Central California and conducts forensic anthropology casework throughout California.

Lisa N. Bright, PhD, is the cultural branch chief district 3, California Department of Transportation. She is a broadly trained anthropologist specializing in mortuary archaeology, bioarchaeology, historical archaeology, and taphonomy.

Kristen A. Broehl is a PhD candidate in the Department of Anthropology at the University of Nevada, Reno. Her research interests include California bioarchaeology, biodistance analysis, and forensic anthropology.

Kasey E. Cole is a PhD student in the Anthropology Department at the University of Utah. Her research uses zooarchaeological, stable isotope, and paleoclimate data to assess the influence of local environments on past human resource acquisition strategies, settlement decisions, interpersonal conflict, and territoriality in western North America.

Maria Cox is a PhD student at the University of Illinois at Urbana-Champaign. She currently serves as assistant editor for *Human Biology* and as a research assistant in the Department of Anthropology. Her research specialization includes the effect of chronic stress and inflammation, vaccine response, and method development.

Jessica Curry is a coroner's investigator for the Butte County Sheriff's Office in northern California. She received her MA in anthropology at California State University, Chico, with a focus on applications of portable X-Ray Fluorescence for resolving commingled remains cases.

Peter Gavette is an archaeologist for the U.S. National Park Service and is the park archaeologist for Golden Gate National Recreation Area in San Francisco, Marin, and San Mateo counties.

Sarah A. Hall is a PhD student at Arizona State University and serves as a graduate teaching associate in the School of Human Evolution and Social Change. She specializes in bioarchaeology, biogeochemistry, and historical archaeology.

Kelsie Hart is the collections manager for the Center for Archaeology Studies at Texas State University, San Marcos.

Angela Locke Barton has an MA in cultural resources management from Sonoma State University. Her work at Golden Gate National Recreation Area's archaeology laboratory has focused on artifact cataloging and conservation, as well as researching historical bottles.

Heather MacInnes graduated with her MA in physical anthropology from California State University, Chico, in 2017. She now is the cultural resources lead and a project manager at Bargas Environmental Consulting based out of Sacramento, California.

Cecily Merwin is an archaeology field technician at Plumas National Forest in Northern California. Her research focuses on zooarchaeological applications for interpreting diet in the prehistoric western United States.

Samuel A. Mijal's research interests include dental anthropology, stable isotope analysis, and 3D modeling.

Colleen F. Milligan is professor of anthropology and codirector of the Human Identification Laboratory at California State University, Chico. She conducts bioarchaeological research on paleopathology associated with skeletal remains from the late nineteenth and early twentieth centuries. She also conducts forensic anthropological casework throughout California.

Kristen E. Pearlstein is the anatomical collections manager at the National Museum of Health and Medicine in Silver Spring, Maryland. She has a PhD in anthropology and has worked with museum collections for 15 years, specializing in the study of trauma and disease in past populations.

Mallory Peters is a PhD student and instructor of record at Binghamton University. Her research focuses on the impact of stress on skeletal metabolism.

Julia Prince-Buitenhuys is an associate environmental planner (archaeologist) at the California Department of Transportation. Her specializations are in human skeletal biology and stable isotope analysis, and she is a coeditor of *Theoretical Approaches in Bioarchaeology.*

Vanessa C. Reeves is a PhD student at the University of South Florida and serves as graduate teaching associate in the Department of Anthropology. She is currently conducting bioarchaeological research in the Transylvanian region of Romania.

Valerie Sgheiza is a PhD student at the University of Illinois at Urbana-Champaign and serves as a course instructor and research assistant in the Department of Anthropology. Her research specializations include age estimation, human variation, and statistical modeling.

Brian F. Spatola is a biological and forensic anthropologist and curator at the National Museum of Health and Medicine in Silver Spring, Maryland. His specializations include human identification, skeletal trauma analysis, paleopathology, and 3D modelling.

Alina Tichinin is a graduate student in anthropology at California State University, Chico.

Karin Wells received her master's in anthropology from California State University, Chico, in 2019. She currently attends Kansas City University College of Medicine.

P. Willey is professor emeritus at California State University, Chico. His previous coedited book (with Douglas D. Scott) is *Health of the Seventh Cavalry: A Medical History.*

Index

Please note that page numbers in *italics* refer to illustrations.

BIOARCHAEOLOGICAL INTERPRETATIONS OF THE HUMAN PAST: LOCAL, REGIONAL, AND GLOBAL PERSPECTIVES

Edited by Clark Spencer Larsen

Ancient Health: Skeletal Indicators of Agricultural and Economic Intensification, edited by Mark Nathan Cohen and Gillian M. M. Crane-Kramer (2007; first paperback edition, 2012)
Bioarchaeology and Identity in the Americas, edited by Kelly J. Knudson and Christopher M. Stojanowski (2009; first paperback edition, 2010)
Island Shores, Distant Pasts: Archaeological and Biological Approaches to the Pre-Columbian Settlement of the Caribbean, edited by Scott M. Fitzpatrick and Ann H. Ross (2010; first paperback edition, 2017)
The Bioarchaeology of the Human Head: Decapitation, Decoration, and Deformation, edited by Michelle Bonogofsky (2011; first paperback edition, 2015)
Bioarchaeology and Climate Change: A View from South Asian Prehistory, by Gwen Robbins Schug (2011; first paperback edition, 2017)
Violence, Ritual, and the Wari Empire: A Social Bioarchaeology of Imperialism in the Ancient Andes, by Tiffiny A. Tung (2012; first paperback edition, 2013)
The Bioarchaeology of Individuals, edited by Ann L. W. Stodder and Ann M. Palkovich (2012; first paperback edition, 2014)
The Bioarchaeology of Violence, edited by Debra L. Martin, Ryan P. Harrod, and Ventura R. Pérez (2012; first paperback edition, 2013)
Bioarchaeology and Behavior: The People of the Ancient Near East, edited by Megan A. Perry (2012; first paperback edition, 2018)
Paleopathology at the Origins of Agriculture, edited by Mark Nathan Cohen and George J. Armelagos (2013)
Bioarchaeology of East Asia: Movement, Contact, Health, edited by Kate Pechenkina and Marc Oxenham (2013)
Mission Cemeteries, Mission Peoples: Historical and Evolutionary Dimensions of Intracemetery Bioarchaeology in Spanish Florida, by Christopher M. Stojanowski (2013)
Tracing Childhood: Bioarchaeological Investigations of Early Lives in Antiquity, edited by Jennifer L. Thompson, Marta P. Alfonso-Durruty, and John J. Crandall (2014)
The Bioarchaeology of Classical Kamarina: Life and Death in Greek Sicily, by Carrie L. Sulosky Weaver (2015)
Victims of Ireland's Great Famine: The Bioarchaeology of Mass Burials at Kilkenny Union Workhouse, by Jonny Geber (2015; first paperback edition, 2018)
Colonized Bodies, Worlds Transformed: Toward a Global Bioarchaeology of Contact and Colonialism, edited by Melissa S. Murphy and Haagen D. Klaus (2017; first paperback edition, 2021)
Bones of Complexity: Bioarchaeological Case Studies of Social Organization and Skeletal Biology, edited by Haagen D. Klaus, Amanda R. Harvey, and Mark N. Cohen (2017)
A World View of Bioculturally Modified Teeth, edited by Scott E. Burnett and Joel D. Irish (2017)
Children and Childhood in Bioarchaeology, edited by Patrick Beauchesne and Sabrina C. Agarwal (2018)
Bioarchaeology of Pre-Columbian Mesoamerica: An Interdisciplinary Approach, edited by Cathy Willermet and Andrea Cucina (2018)
Massacres: Bioarchaeology and Forensic Anthropology Approaches, edited by Cheryl P. Anderson and Debra L. Martin (2019)

Mortuary and Bioarchaeological Perspectives on Bronze Age Arabia, edited by Kimberly D. Williams and Lesley A. Gregoricka (2019)

Bioarchaeology of Frontiers and Borderlands, edited by Cristina I. Tica and Debra L. Martin (2019)

The Odd, the Unusual, and the Strange: Bioarchaeological Explorations of Atypical Burials, edited by Tracy K. Betsinger, Amy B. Scott, and Anastasia Tsaliki (2020)

Bioarchaeology and Identity Revisited, edited by Kelly J. Knudson and Christopher M. Stojanowski (2020)

Leprosy: Past and Present, by Charlotte A. Roberts (2020)

Bioarchaeology of Care through Population-Level Analyses, edited by Alecia Schrenk and Lori A. Tremblay (2022)

Archaeology and Bioarchaeology of Anatomical Dissection at a Nineteenth-Century Army Hospital in San Francisco, edited by P. Willey, Peter Gavette, Eric J. Bartelink, and Colleen F. Milligan (2023)

www.ingramcontent.com/pod-product-compliance
Lightning Source LLC
LaVergne TN
LVHW011806200926
848484LV00005B/1086